EVERYDAY Literacy
Math

GRADE 1

Download Home–School Activities in Spanish

The Home–School Connection at the end of each weekly lesson in the book is also available in Spanish on our website.

How to Download:

1. Go to www.evan-moor.com/resources.
2. Enter your e-mail address and the resource code for this product—EMC3039.
3. You will receive an e-mail with a link to the downloadable letters, as well as an attachment with instructions.

Editorial Development: Camille Liscinsky
Guadalupe Lopez
Lisa Vitarisi Mathews
Copy Editing: Anna Pelligra
Art Direction: Kathy Kopp
Cover Design: Marcia Bateman
Illustration: Cheryl Nobens
Design/Production: Carolina Caird
Jessica Onken

EMC 3039
Evan-Moor
EDUCATIONAL PUBLISHERS
Helping Children Learn since 1979

Visit
teaching-standards.com
to view a correlation
of this book.
This is a free service.

Correlated to State and Common Core State Standards

Congratulations on your purchase of some of the finest teaching materials in the world.

Photocopying the pages in this book is permitted for single-classroom use only. Making photocopies for additional classes or schools is prohibited.

For information about other Evan-Moor products, call 1-800-777-4362, fax 1-800-777-4332, or visit our Web site, www.evan-moor.com.
Entire contents © 2013 EVAN-MOOR CORP.
18 Lower Ragsdale Drive, Monterey, CA 93940-5746. Printed in USA.

CPSIA: Printed by McNaughton & Gunn, Saline, MI USA. [11/2012]

Contents

What's Inside 4
How to Use This Book 5
Skills Chart 6
Student Progress Record 8
Small-Group Record Sheet 9
Letter to Parents 10

Week 1: **How Many?** *(Number Sense)* 11
Week 2: **Number Names** *(Number Sense)* 19
Week 3: **Count by 10s** *(Number Sense)* 27
Week 4: **Skip Counting** *(Number Sense)* 35
Week 5: **Tens and Ones** *(Number Sense)* 43
Week 6: **How Many In All?** *(Algebra)* 51
Week 7: **How Many Are Left?** *(Algebra)* 59
Week 8: **Order of Addends** *(Algebra)* 67
Week 9: **Fact Families** *(Algebra)* 75
Week 10: **What's the Problem?** *(Algebra)* 83
Week 11: **Greater Than, Less Than** *(Algebra)* 91
Week 12: **Basic Shapes** *(Geometry)* 99
Week 13: **Fractions** *(Geometry)* 107
Week 14: **Ways to Measure** *(Measurement)* 115
Week 15: **Measure with Inches** *(Measurement)* . . . 123
Week 16: **Time for Fun!** *(Measurement)* 131
Week 17: **Patterns** *(Algebra)* 139
Week 18: **Where Is It?** *(Geometry)* 147
Week 19: **Counting Coins** *(Number Sense)* 155
Week 20: **Using Graphs** *(Data Analysis)* 163

Answer Key 171

© Evan-Moor Corp. • EMC 3039 • Everyday Literacy: Math

What's Inside

In this book, you will find **20 weekly lessons**. Each weekly lesson includes:

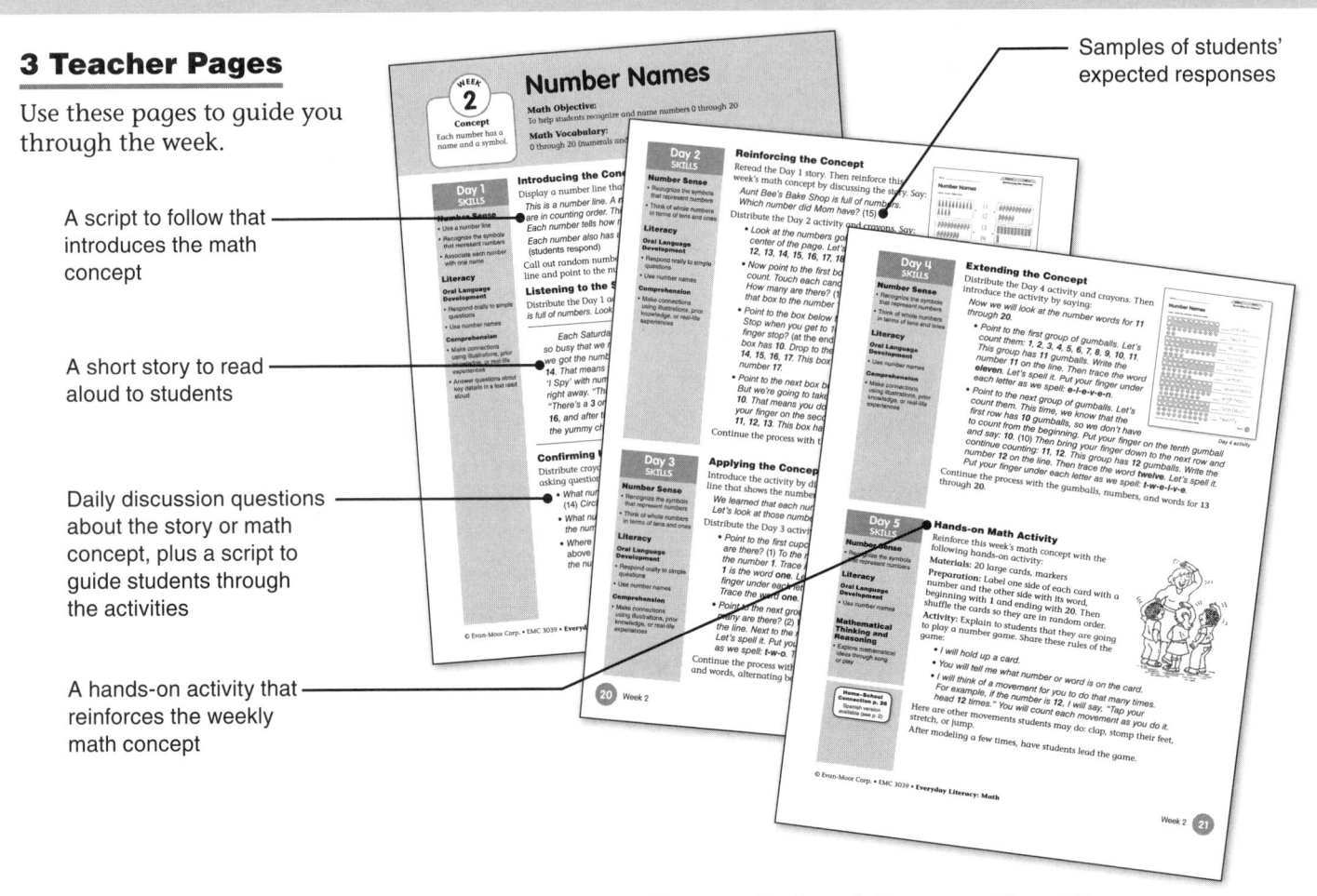

3 Teacher Pages

Use these pages to guide you through the week.

- A script to follow that introduces the math concept
- A short story to read aloud to students
- Daily discussion questions about the story or math concept, plus a script to guide students through the activities
- A hands-on activity that reinforces the weekly math concept

Samples of students' expected responses

4 Student Activity Pages

Reproduce each page for students to complete during the daily lesson.

1 Home–School Connection Page

At the end of each week, give students the **Home–School Connection** page (in English or Spanish) to take home and share with their parents.

To access the Spanish version of the page, go to www.evan-moor.com/resources. Enter your e-mail address and the resource code EMC3039.

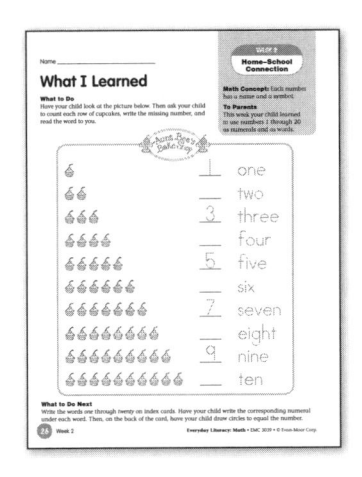

Everyday Literacy: Math • EMC 3039 • © Evan-Moor Corp.

How to Use This Book

Follow these easy steps to conduct the lessons:

Day 1

Reproduce and distribute the *Day 1 Student Page* to each student.

Use the scripted *Day 1 Teacher Page* to:

1. Introduce the weekly concept.
2. Read the story aloud as students listen and look at the picture.
3. Guide students through the activity.

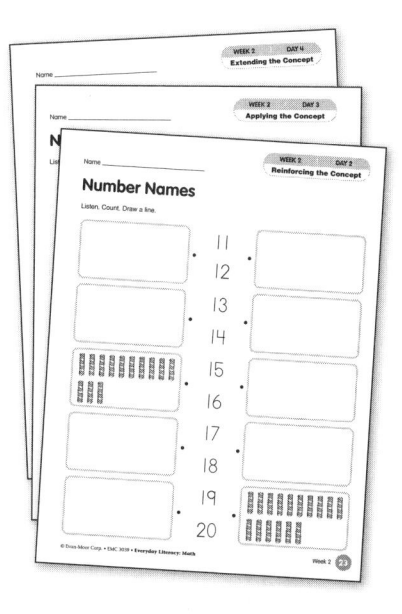

Days 2, 3, and 4

Reproduce and distribute the appropriate day's activity page to each student.

Use the scripted *Teacher Page* to:

1. Review and discuss the Day 1 story.
2. Reinforce, apply, or extend the math concept.
3. Guide students through the activity.

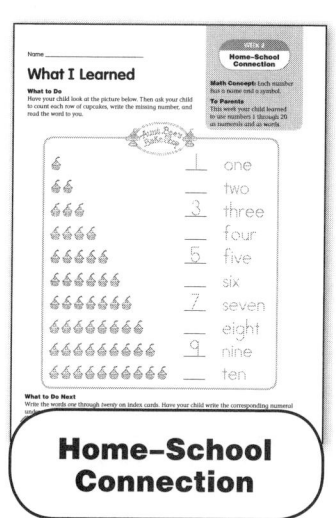

Day 5

Follow the directions to lead the *Hands-on Math Activity*.

Send home the **Home–School Connection** page for each student to complete with his or her parents.

Tips for Success

- Review the *Teacher Page* before you begin the lesson.
- Work with students in small groups at a table in a quiet area of the room.
- Model how to respond to questions by using complete sentences. For example, if a student responds to the question "What number is on the cake?" by answering "3," you'd respond, "That's right. The number 3 is on the cake."
- Wait for students to complete each task before giving the next direction.
- Provide visual aids or concrete demonstrations when possible.

© Evan-Moor Corp. • EMC 3039 • *Everyday Literacy: Math*

5

Skills Chart

	Math																	
	Number Sense									Data Analysis	Algebra							
Week	Use a hundred chart	Use a number line	Apply one-to-one correspondence when counting	Understand the relationship between numbers and quantities	Understand that the final number counted in a set tells how many	Recognize the symbols that represent numbers	Think of whole numbers in terms of tens and ones	Determine the value of a group of coins	Count by 2s, 5s, or 10s	Associate each number with one name	Organize, record, and interpret data	Use more than one strategy to solve a problem	Determine the unknown whole number in an equation or word problem	Understand the commutative property of addition	Solve addition problems	Understand the relationship between addition and subtraction	Compare numbers as *greater than*, *less than*, or *equal to*	Use symbols that indicate a comparison	Recognize and form simple patterns
1			•	•	•														
2		•	•	•		•	•		•										
3	•				•				•										
4		•	•		•				•										
5			•				•		•										
6												•	•						
7												•	•						
8														•	•				
9																•			
10												•							
11																	•	•	
12																			
13																			
14																			
15																			
16																			
17										•									•
18																			
19								•	•										
20											•								

Skills Scope and Sequence

	Geometry				Measurement			Mathematical Thinking and Reasoning					Oral Language Development						Comprehension			
	Recognize simple shapes regardless of size or orientation	Understand that shapes can be divided into equal parts called halves or fourths	Understand that decomposing a shape results in smaller parts	Find locations on a grid	Use a nonstandard unit to measure	Use a ruler to measure	Tell time to the hour and half-hour	Explore mathematical ideas through song or play	Select and use various types of reasoning and methods of proof	Use math to solve problems	Use number concepts for a meaningful purpose	Record observations and data with pictures and other symbols	Respond orally to simple questions	Use number names	Use mathematical terms	Use the names of shapes	Use measurement terms	Use vocabulary related to time concepts	Make connections using illustrations, prior knowledge, or real-life experiences	Answer questions about key details in a text read aloud	Make inferences and draw conclusions	Week
								•					•	•					•	•		1
								•					•	•					•	•		2
								•					•	•					•	•		3
								•					•	•					•	•		4
									•				•	•	•				•	•		5
								•					•		•				•	•		6
									•				•		•				•	•		7
								•					•		•				•	•		8
										•			•		•				•	•		9
								•					•		•				•	•		10
									•				•		•				•	•		11
	•							•					•			•			•	•		12
		•	•						•				•		•				•	•	•	13
					•					•			•				•		•	•		14
						•					•		•				•		•	•		15
							•			•			•		•			•	•	•		16
								•					•		•				•			17
			•					•					•		•				•	•		18
											•		•		•				•	•		19
											•		•		•				•	•	•	20

© Evan-Moor Corp. • EMC 3039 • **Everyday Literacy: Math**

Use this record sheet periodically to check a student's progress.

Everyday Literacy
Math

1

Student Progress Record

Name: _____

Write dates and comments in the boxes below the student's proficiency level.

1: Rarely demonstrates   0 – 25 %
2: Occasionally demonstrates   25 – 50 %
3: Usually demonstrates   50 – 75 %
4: Consistently demonstrates   75 – 100 %

Literacy Concepts

	1	2	3	4
Writes numbers legibly				
Writes equations from left to right and top to bottom				
Understands that each number is represented by a numerical symbol and a word				

Oral Language Development

Uses descriptive language				
Counts pictured objects aloud				
Responds orally to simple questions				

Comprehension

Makes connections using illustrations, prior knowledge, or real-life experiences				
Answers questions about key details in a text read aloud				

Math

Uses mathematical terms when speaking				
Engages in mathematical thinking and reasoning				
Applies mathematical concepts in written practice activities				

Everyday Literacy: Math • EMC 3039 • © Evan-Moor Corp.

Everyday Literacy
Math

Small-Group Record Sheet

Students' Names:

Write dates and comments about students' performance each week.

Week	Title	Comments
1	How Many?	
2	Number Names	
3	Count by 10s	
4	Skip Counting	
5	Tens and Ones	
6	How Many In All?	
7	How Many Are Left?	
8	Order of Addends	
9	Fact Families	
10	What's the Problem?	
11	Greater Than, Less Than	
12	Basic Shapes	
13	Fractions	
14	Ways to Measure	
15	Measure with Inches	
16	Time for Fun	
17	Patterns	
18	Where Is It?	
19	Counting Coins	
20	Using Graphs	

© Evan-Moor Corp. • EMC 3039 • **Everyday Literacy: Math**

Dear Parent or Guardian,

Every week your child will learn a concept related to number sense, geometry, data analysis, measurement, or algebra. Your child will develop oral language and comprehension skills by listening to stories and engaging in oral, written, and hands-on activities that reinforce math concepts.

At the end of each week, I will send home an activity page for you to complete with your child. The activity page reviews the weekly math concept and has an activity for you and your child to do together.

Sincerely,

Estimado padre o tutor:

Cada semana su niño(a) aprenderá sobre un concepto de matemáticas relacionado a la noción de los números, geometría, análisis de datos, medidas o álgebra. Su niño(a) desarrollará las habilidades de lenguaje oral y de comprensión escuchando cuentos y realizando actividades orales y escritas. Además, participará en actividades prácticas que apoyan los conceptos de matemáticas.

Al final de cada semana, le enviaré una hoja de actividades para que la complete en casa con su niño(a). La hoja repasa el concepto de matemáticas de la semana, y contiene una actividad que pueden completar usted y su niño(a) juntos.

Atentamente,

Everyday Literacy: Math • EMC 3039 • © Evan-Moor Corp.

WEEK 1

Concept
Counting tells how many are in a set.

How Many?

Math Objective:
To help students review counting to tell how many

Math Vocabulary:
altogether, count, each

Day 1 SKILLS

Number Sense
- Apply one-to-one correspondence when counting
- Understand the relationship between numbers and quantities
- Understand that the final number counted in a set tells how many

Literacy

Oral Language Development
- Respond orally to simple questions
- Use number names

Comprehension
- Make connections using illustrations, prior knowledge, or real-life experiences
- Answer questions about key details in a text read aloud

Introducing the Concept

Display a set of 10 blocks and a set of 8 blocks. Model counting each set of blocks, touching each block as you count. Then ask:

- *How did I find out how many blocks are in each group?* (count)
- *Which group has 8 blocks?* (students respond) *Let's count this set together: 1, 2, 3, 4, 5, 6, 7, 8.*
- *Which group has 10 blocks?* (students respond) *Let's count this set together: 1, 2, 3, 4, 5, 6, 7, 8, 9, 10.*

Listening to the Story

Distribute the Day 1 activity. Say: *Look at the picture and listen as I read a story about animals in a coconut tree.*

In the middle of the jungle, there is a big coconut tree. Different kinds of animals live and play in that tree. **Four** monkeys run up and down the tree all day. **Three** birds, called toucans, are up high, waving their big beaks. **Five** graceful butterflies flutter about. And **six** fuzzy coconuts hang from the tree, ready to fall at any time. The coconut tree is the place to be!

Confirming Understanding

Distribute crayons. Develop the math concept by asking questions about the story. Ask:

- *How many monkeys are in the coconut tree?* (4) *Trace the number 4 by the monkeys.*
- *How many coconuts are hanging from the tree?* (6) *Trace the number 6 by the coconuts.*
- *How many toucans are in the tree?* (3) *Trace the number 3 next to the toucans.*
- *How many butterflies are flying about?* (5) *Trace the number 5 next to the butterflies.*

Day 1 picture

© Evan-Moor Corp. • EMC 3039 • Everyday Literacy: Math

Week 1

Day 2 SKILLS

Number Sense
- Apply one-to-one correspondence when counting
- Understand the relationship between numbers and quantities
- Understand that the final number counted in a set tells how many

Literacy

Oral Language Development
- Respond orally to simple questions
- Use number names

Comprehension
- Make connections using illustrations, prior knowledge, or real-life experiences

Reinforcing the Concept

Reread the Day 1 story. Then reinforce this week's math concept by discussing the story. Ask:

Were there more monkeys or more toucans in the coconut tree? (monkeys)

Distribute the Day 2 activity and crayons. Say:

- *Point to box 1. Here are some coconut trees. How many coconut trees are there? Point to each one as you count.* (2) *That's right. Circle the 2 below the coconut trees.*

- *Point to box 2. Here are some coconuts. How many coconuts are there? Point to each one as you count.* (8) *That's right. Circle the 8 below the coconuts.*

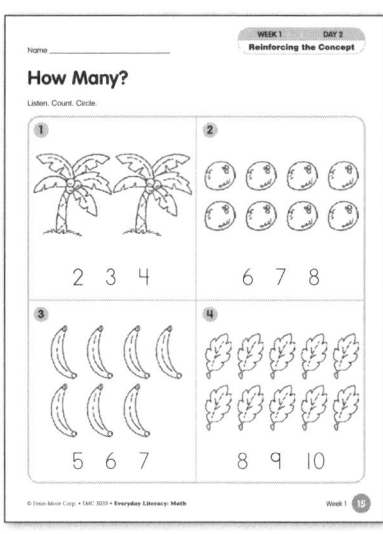

- *Point to box 3. Here are some bananas. How many bananas are there? Point to each one as you count.* (7) *That's right. Circle the 7 below the bananas.*

- *Point to box 4. Here are some leaves. How many leaves are there? Point to each one as you count.* (10) *That's right. Circle the 10 below the leaves.*

Day 2 activity

Day 3 SKILLS

Number Sense
- Apply one-to-one correspondence when counting
- Understand the relationship between numbers and quantities
- Understand that the final number counted in a set tells how many

Literacy

Oral Language Development
- Respond orally to simple questions
- Use number names

Comprehension
- Make connections using illustrations, prior knowledge, or real-life experiences

Applying the Concept

Introduce the activity by presenting a number line that includes the numbers **0** through **20**. Lead students in counting from **1** to **20**, pointing to each number as you name it. Say:

*So far you have counted items **fewer** than 10. Now you will count **higher** than 10. You will count things up to 20. Look at the number line and let's count together from 1 to 20.*

Distribute the Day 3 activity and crayons. Say:

- *Point to the box with ants. How many ants are there?* (12) *That's right. Draw a line from the ants to the number **12**. Trace the number **12**.*

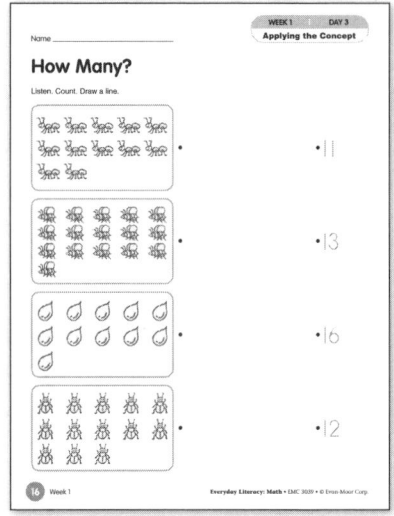

- *Point to the box with bees. How many bees are there?* (16) *That's right. Draw a line from the bees to the number **16**. Trace the number **16**.*

- *Point to the box with raindrops. How many raindrops are there?* (11) *That's right. Draw a line from the raindrops to the number **11**. Trace the number **11**.*

- *Point to the box with beetles. How many beetles are there?* (13) *That's right. Draw a line from the beetles to the number **13**. Trace the number **13**.*

Day 3 activity

Day 4 SKILLS

Number Sense
- Apply one-to-one correspondence when counting
- Understand the relationship between numbers and quantities
- Understand that the final number counted in a set tells how many

Literacy

Oral Language Development
- Respond orally to simple questions
- Use number names

Comprehension
- Make connections using illustrations, prior knowledge, or real-life experiences

Extending the Concept

Distribute the Day 4 activity and crayons. Guide students through the activity by saying:

- *Point to box 1. Here are some mangoes. How many mangoes are there? Point to each one as you count. (13) That's right. Write the number **13** on the line below the mangoes.*

- *Point to box 2. Here are some spiders. How many spiders are there? Point to each one as you count. (11) That's right. Write the number **11** on the line below the spiders.*

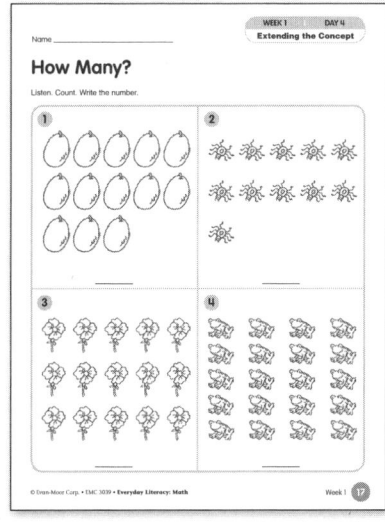

- *Point to box 3. Here are some flowers. How many flowers are there? Point to each one as you count. (15) That's right. Write the number **15** on the line below the flowers.*

- *Point to box 4. Here are some frogs. How many frogs are there? Point to each one as you count. (20) That's right. Write the number **20** on the line below the frogs.*

Day 4 activity

Day 5 SKILLS

Number Sense
- Understand the relationship between numbers and quantities

Literacy

Oral Language Development
- Use number names

Mathematical Thinking and Reasoning
- Explore mathematical ideas through song or play

Home–School Connection p. 18
Spanish version available (see p. 2)

Circle Time Math Activity

Reinforce this week's math concept with the following circle time activity:

Materials: index cards, dot stickers, string

Preparation: Make two sets of cards: number one set **1** through **20** and draw corresponding dot arrays **1** through **20** on the other set. Attach string to make each card into a necklace for students to wear.

Activity: Have students stand in a circle. Give one necklace to each student to wear. Have each student look at his or her necklace and identify the number on it or the number the array stands for. Then explain how to play "Number Moves."

How to Play: To start the game, call out a random number from **1** through **20**. For example, call out the number **12**. The two students who have cards that represent the number **12** come to the center of the circle. Have the students check each other's cards to make sure they both represent **12**. If they do, then those two students are the leaders of the first round of "Number Moves." Have the two collaborate to think of **12** repetitive movements they can ask the rest of the class to do. For example, they may say, *Raise your hands above your head and clap 12 times.*

Then have those leaders return to the circle. Continue the game by calling out a different number.

© Evan-Moor Corp. • EMC 3039 • **Everyday Literacy: Math** Week 1 13

Name _____

WEEK 1 | DAY 1
Confirming Understanding

How Many?

3

6

5

4

14 Week 1

Everyday Literacy: Math • EMC 3039 • © Evan-Moor Corp.

Name _____

WEEK 1 | DAY 2
Reinforcing the Concept

How Many?

Listen. Count. Circle.

1

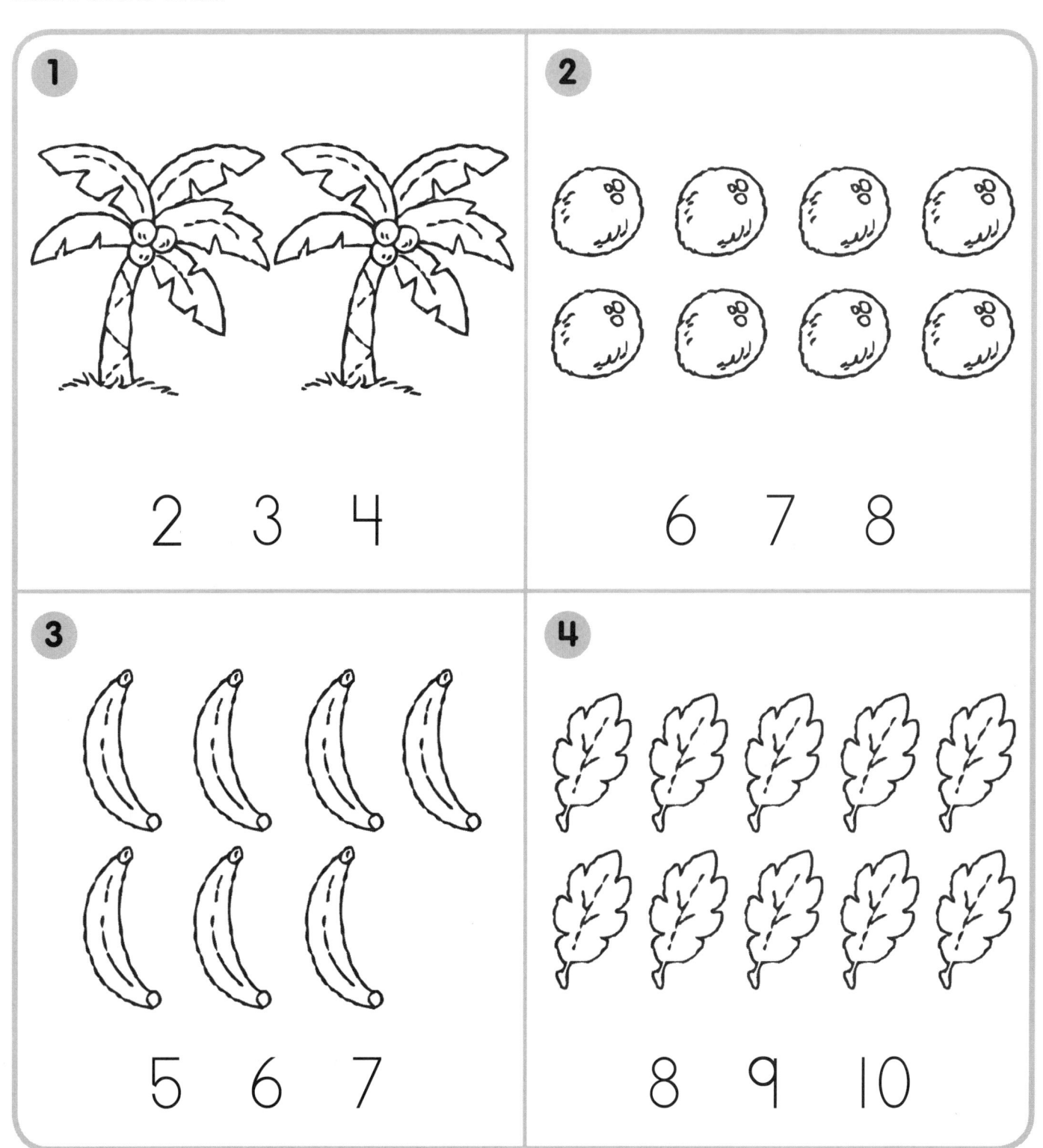

2   3   4

2

6   7   8

3

5   6   7

4

8   9   10

© Evan-Moor Corp. • EMC 3039 • **Everyday Literacy: Math**   Week 1   15

Name _____

WEEK 1 | DAY 3
Applying the Concept

How Many?

Listen. Count. Draw a line.

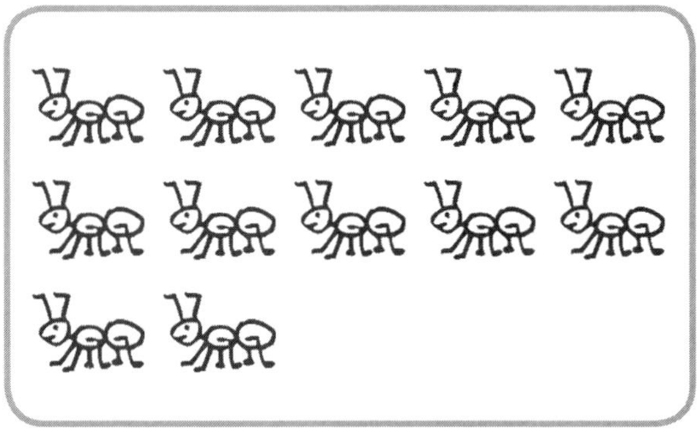

• 11

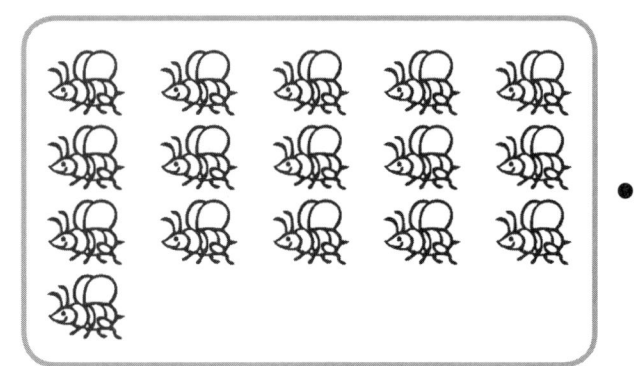

• 13

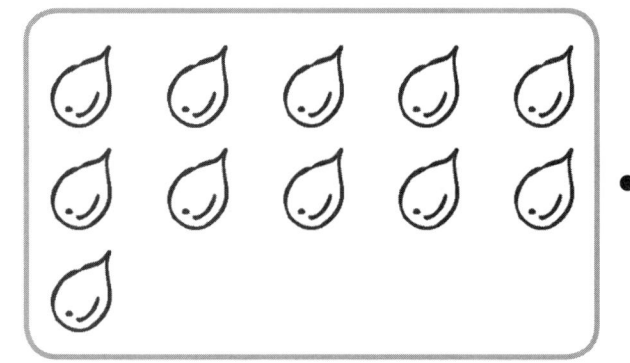

• 16

• 12

16 Week 1

Everyday Literacy: Math • EMC 3039 • © Evan-Moor Corp.

Name _____

WEEK 1 | DAY 4
Extending the Concept

How Many?

Listen. Count. Write the number.

1

_____

2

_____

3

_____

4

_____

© Evan-Moor Corp. • EMC 3039 • **Everyday Literacy: Math**

Week 1   17

Name _____

What I Learned

What to Do
Have your child look at the picture below and tell you how many animals and coconuts are in the coconut tree. Then have your child trace each number.

WEEK 1

Home–School Connection

Math Concept: Counting tells how many are in a set.

To Parents
This week your child learned to count items up to 20.

What to Do Next
Display groups of items up to 20. Give your child a pad of sticky notes and have him or her write the number of items in each group and then place the note near the group.

Week 1   Everyday Literacy: Math • EMC 3039 • © Evan-Moor Corp.

WEEK 2

Concept
Each number has a name and a symbol.

Number Names

Math Objective:
To help students recognize and name numbers 0 through 20

Math Vocabulary:
0 through 20 (numerals and words), number line

Day 1 SKILLS

Number Sense
- Use a number line
- Recognize the symbols that represent numbers
- Associate each number with one name

Literacy

Oral Language Development
- Respond orally to simple questions
- Use number names

Comprehension
- Make connections using illustrations, prior knowledge, or real-life experiences
- Answer questions about key details in a text read aloud

Introducing the Concept

Display a number line that begins at **0** and ends at **20**. Say:

*This is a number line. A **number line** has numbers on it. The numbers are in counting order. This number line shows the numbers **0** through **20**. Each number tells how many.*

Each number also has a name. Name the numbers with me. (students respond)

Call out random numbers and have volunteers come up to the number line and point to the number.

Listening to the Story

Distribute the Day 1 activity. Say: *I will read a story about a bake shop that is full of numbers. Look at the picture and listen as I read.*

---

*Each Saturday, Mom and I visit Aunt Bee's Bake Shop. It is always so busy that we need to take a number and wait for our turn. Today, we got the number **15**. Aunt Bee is serving a lady who has the number **14**. That means we are next, but we must wait. Mom says, "Let's play 'I Spy' with numbers. I'll start. I spy a chocolate **0**." I see the yummy **0** right away. "That doughnut is shaped like a **0**! I spy a number **3**." "There's a **3** on that cake," says Mom. After that, we spot a **12** and a **16**, and after that, it's our turn! Mom gets a cherry cupcake and I get the yummy chocolate doughnut in the shape of a **0**.*

Confirming Understanding

Distribute crayons. Develop the math concept by asking questions about the story. Ask:

- *What number is the ladybug holding?* (14) *Circle the number **14**.*
- *What number is on the cake?* (3) *Trace the number **3** in blue.*
- *Where is the number **12**?* (on a sign above the cookies) *Draw a line under the number **12**.*

Day 1 picture

Week 2

Day 2
SKILLS

Number Sense
- Recognize the symbols that represent numbers
- Think of whole numbers in terms of tens and ones

Literacy

Oral Language Development
- Respond orally to simple questions
- Use number names

Comprehension
- Make connections using illustrations, prior knowledge, or real-life experiences

Reinforcing the Concept

Reread the Day 1 story. Then reinforce this week's math concept by discussing the story. Say:

Aunt Bee's Bake Shop is full of numbers. Which number did Mom have? (15)

Distribute the Day 2 activity and crayons. Say:

- *Look at the numbers going down the center of the page. Let's count them:* **11, 12, 13, 14, 15, 16, 17, 18, 19, 20.**

- *Now point to the first box of candles. Let's count. Touch each candle as we count. How many are there?* (14) *Draw a line from that box to the number* **14**.

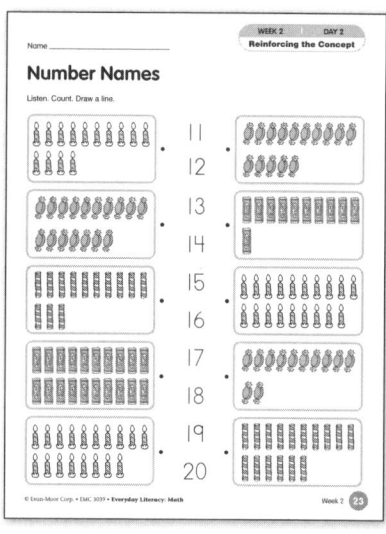

- *Point to the box below that one. Let's count the candies. Stop when you get to 10:* **1, 2, 3, 4, 5, 6, 7, 8, 9, 10.** *Where did your finger stop?* (at the end of the row) *That's right. The first row of each box has* **10**. *Drop to the next row and let's keep counting:* **11, 12, 13, 14, 15, 16, 17.** *This box has* **17** *candies. Draw a line from this box to the number* **17**.

- *Point to the next box below. We're going to count these sticks of candy. But we're going to take a shortcut. Remember: The first row always has* **10**. *That means you don't have to count from the very beginning. Put your finger on the second row. Count, starting with the number after 10:* **11, 12, 13.** *This box has* **13** *sticks. Draw a line to the number* **13**.

Continue the process with the rest of the boxes.

Day 2 activity

Day 3
SKILLS

Number Sense
- Recognize the symbols that represent numbers
- Think of whole numbers in terms of tens and ones

Literacy

Oral Language Development
- Respond orally to simple questions
- Use number names

Comprehension
- Make connections using illustrations, prior knowledge, or real-life experiences

Applying the Concept

Introduce the activity by displaying a number line that shows the numbers **1** through **20**. Say:

We learned that each number has a name. Let's look at those number words.

Distribute the Day 3 activity and crayons. Say:

- *Point to the first cupcake. How many are there?* (1) *To the right of that cupcake is the number* **1**. *Trace it. Next to the number* **1** *is the word* **one**. *Let's spell it. Put your finger under each letter as we spell:* **o-n-e**. *Trace the word* **one**.

- *Point to the next group of cupcakes. How many are there?* (2) *Write the number* **2** *on the line. Next to the number* **2** *is the word* **two**. *Let's spell it. Put your finger under each letter as we spell:* **t-w-o**. *Trace the word* **two**.

Continue the process with the remaining rows of the cupcakes, numbers, and words, alternating between tracing and writing the numbers.

Day 3 activity

20 Week 2   Everyday Literacy: Math • EMC 3039 • © Evan-Moor Corp.

Day 4 SKILLS

Number Sense
- Recognize the symbols that represent numbers
- Think of whole numbers in terms of tens and ones

Literacy

Oral Language Development
- Use number names

Comprehension
- Make connections using illustrations, prior knowledge, or real-life experiences

Extending the Concept

Distribute the Day 4 activity and crayons. Then introduce the activity by saying:

*Now we will look at the number words for **11** through **20**.*

- *Point to the first group of gumballs. Let's count them: **1, 2, 3, 4, 5, 6, 7, 8, 9, 10, 11**. This group has **11** gumballs. Write the number **11** on the line. Then trace the word **eleven**. Let's spell it. Put your finger under each letter as we spell: **e-l-e-v-e-n**.*

- *Point to the next group of gumballs. Let's count them. This time, we know that the first row has **10** gumballs, so we don't have to count from the beginning. Put your finger on the tenth gumball and say: **10**. (10) Then bring your finger down to the next row and continue counting: **11, 12**. This group has **12** gumballs. Write the number **12** on the line. Then trace the word **twelve**. Let's spell it. Put your finger under each letter as we spell: **t-w-e-l-v-e**.*

Continue the process with the gumballs, numbers, and words for **13** through **20**.

Day 4 activity

Day 5 SKILLS

Number Sense
- Recognize the symbols that represent numbers

Literacy

Oral Language Development
- Use number names

Mathematical Thinking and Reasoning
- Explore mathematical ideas through song or play

Home–School Connection p. 26
Spanish version available (see p. 2)

Hands-on Math Activity

Reinforce this week's math concept with the following hands-on activity:

Materials: 20 large cards, markers

Preparation: Label one side of each card with a number and the other side with its word, beginning with **1** and ending with **20**. Then shuffle the cards so they are in random order.

Activity: Explain to students that they are going to play a number game. Share these rules of the game:

- *I will hold up a card.*
- *You will tell me what number or word is on the card.*
- *I will think of a movement for you to do that many times. For example, if the number is **12**, I will say, "Tap your head **12** times." You will count each movement as you do it.*

Here are other movements students may do: clap, stomp their feet, stretch, or jump.

After modeling a few times, have students lead the game.

© Evan-Moor Corp. • EMC 3039 • Everyday Literacy: Math

Week 2 21

Name _____

WEEK 2 | DAY 1
Confirming Understanding

Number Names

Name _____

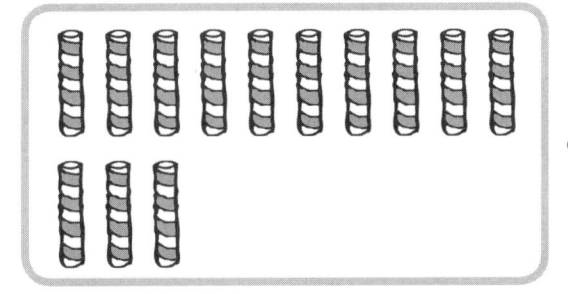

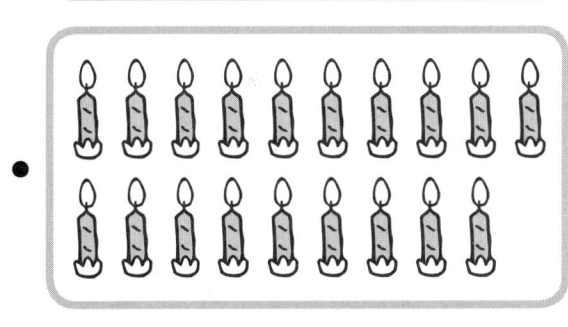

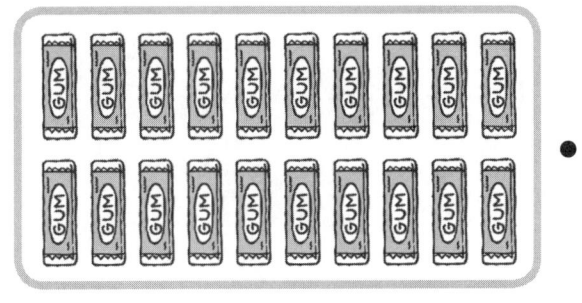

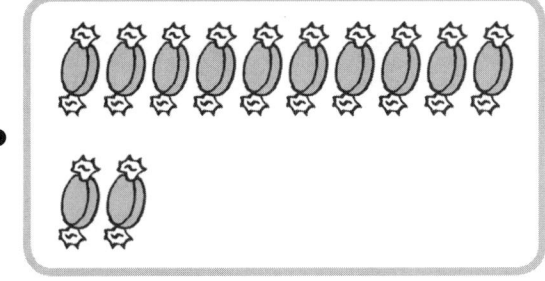

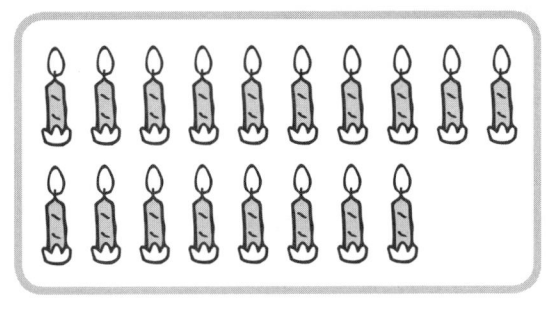

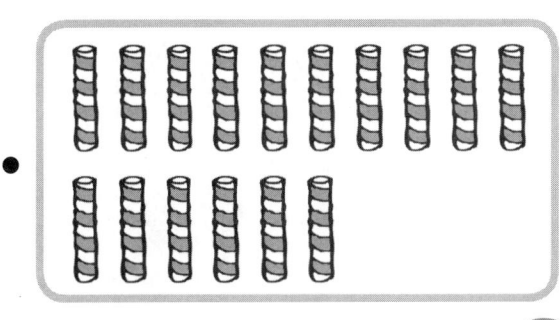

WEEK 2 | DAY 2
Reinforcing the Concept

Number Names

Listen. Count. Draw a line.

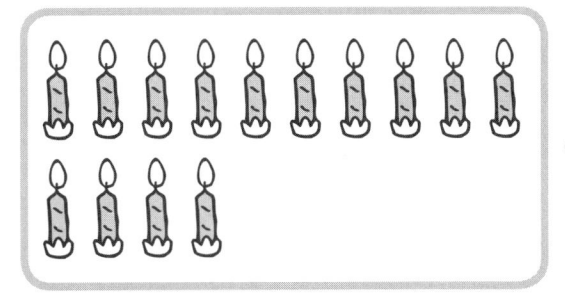

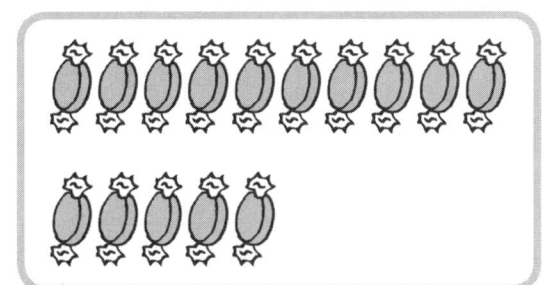

11
12
13
14
15
16
17
18
19
20

Week 2 23

Name _____

WEEK 2 | DAY 3
Applying the Concept

Number Names

Listen. Write or trace the numbers. Trace the words.

Aunt Bee's Bake Shop

🧁	1	one
🧁🧁	__	two
🧁🧁🧁	3	three
🧁🧁🧁🧁	__	four
🧁🧁🧁🧁🧁	5	five
🧁🧁🧁🧁🧁🧁	__	six
🧁🧁🧁🧁🧁🧁🧁	7	seven
🧁🧁🧁🧁🧁🧁🧁🧁	__	eight
🧁🧁🧁🧁🧁🧁🧁🧁🧁	9	nine
🧁🧁🧁🧁🧁🧁🧁🧁🧁🧁	__	ten

Name _____

WEEK 2 | DAY 4
Extending the Concept

Number Names

Listen. Write the numbers. Trace the words.

___ eleven

___ twelve

___ thirteen

___ fourteen

___ fifteen

___ sixteen

___ seventeen

___ eighteen

___ nineteen

___ twenty

© Evan-Moor Corp. • EMC 3039 • Everyday Literacy: Math

Week 2

Name _____

What I Learned

What to Do
Have your child look at the picture below. Then ask your child to count each row of cupcakes, write the missing number, and read the word to you.

WEEK 2

Home–School Connection

Math Concept: Each number has a name and a symbol.

To Parents
This week your child learned to use numbers 1 through 20 as numerals and as words.

Aunt Bee's Bake Shop

Cupcakes	Numeral	Word
🧁	1	one
🧁🧁	___	two
🧁🧁🧁	3	three
🧁🧁🧁🧁	___	four
🧁🧁🧁🧁🧁	5	five
🧁🧁🧁🧁🧁🧁	___	six
🧁🧁🧁🧁🧁🧁🧁	7	seven
🧁🧁🧁🧁🧁🧁🧁🧁	___	eight
🧁🧁🧁🧁🧁🧁🧁🧁🧁	9	nine
🧁🧁🧁🧁🧁🧁🧁🧁🧁🧁	___	ten

What to Do Next
Write the words *one* through *twenty* on index cards. Have your child write the corresponding numeral under each word. Then, on the back of the card, have your child draw circles to equal the number.

Week 2   Everyday Literacy: Math • EMC 3039 • © Evan-Moor Corp.

WEEK 3

Concept
The number system is based on 10.

Count by 10s

Math Objective:
To help students skip count by tens to 100

Math Vocabulary:
count by 10s, hundred chart, ones, skip count, tens

Day 1 SKILLS

Number Sense
- Use a hundred chart
- Understand that the final number counted in a set tells how many
- Count by 2s, 5s, or 10s

Literacy

Oral Language Development
- Respond orally to simple questions
- Use number names

Comprehension
- Make connections using illustrations, prior knowledge, or real-life experiences
- Answer questions about key details in a text read aloud

Introducing the Concept

To prepare for the lesson, have a hundred chart and 3 towers of 10-count linker cubes on hand. Say:

*We're going to count to **100** on this chart. We call this a **hundred chart**. It has numbers from **1** to **100**. We can count from **1** to **100** by counting each number one at a time. But that would take a long time! A faster way to count to 100 is to count by groups of ten. When we count by **tens**, we **skip count**. Listen to me count by **tens** as I point to each number: **10, 20, 30, 40, 50, 60, 70, 80, 90, 100**.*

Have students echo you as you count by tens again. Then call students' attention to the 3 towers of cubes. Say:

*I wonder how many cubes I have in all. I have three groups of **10**. In other words, I have **3 tens**. As I point to each tower, let's count by tens: **10, 20, 30**. How many cubes are there in all? (30) Counting by tens is a much faster way to count!*

Listening to the Story

Distribute the Day 1 activity. Say: *Look at the picture and listen as I read a story about a girl who counts by tens.*

---

*Uncle Finn, Lucy's favorite uncle, loves fish! He has **5** fish tanks in his house. One day, Lucy asked her uncle how many fish he had. "I'll give you a clue," smiled Uncle Finn. "Each tank has **10** fish." "Ah!" said Lucy. She walked to each fish tank and counted by tens: "**10, 20, 30, 40, 50**. You have **50** fish!" "Yes," said Uncle Finn. "**Fifty** nifty fish!"*

Confirming Understanding

Distribute crayons. Develop the math concept by asking questions about the story. Ask:

- *How many fish were in each tank?* (10)
- *Count each tank. Draw a dot on each tank as you count. How many tanks are there?* (5) *Yes. There are **5** groups of **10**. There are **5 tens**.*

Day 1 picture

© Evan-Moor Corp. • EMC 3039 • Everyday Literacy: Math

Week 3   27

Day 2 SKILLS

Number Sense
- Count by 2s, 5s, or 10s
- Understand that the final number counted in a set tells how many

Literacy

Oral Language Development
- Respond orally to simple questions
- Use number names

Comprehension
- Make connections using illustrations, prior knowledge, or real-life experiences
- Answer questions about key details in a text read aloud

Reinforcing the Concept

Reread the Day 1 story. Then reinforce this week's math concept by discussing the story. Say:

*Uncle Finn had **5** fish tanks. Each tank had **10** fish. How many fish did he have in all?* (50)

Distribute the Day 2 activity and crayons. Say:

- *Point to number 1. Let's count the fish. First, let's count by ones: **1, 2, 3, 4, 5, 6, 7, 8, 9, 10**. How many fish are there?* (10) *There are **10** fish. Trace the word **ten**. Now write the number **10** in the box.*

- *Point to number 2. Let's count the fish again. Let's count by ones.* (students count) *Now let's count by tens. Point to each group of **10** fish as we count: **10, 20**. That was much faster! How many fish are there?* (20) *Trace the word **twenty**. Then write the number **20** in the box.*

- *Point to number 3. Let's count the fish one last time. There are lots of fish, and we know that each group has **10** fish. Let's save some time and count by tens. Point to each group as we count: **10, 20, 30**. How many fish are there?* (30) *Trace the word **thirty**. Then write the number **30** in the box.*

Day 2 activity

Day 3 SKILLS

Number Sense
- Count by 2s, 5s, or 10s
- Understand that the final number counted in a set tells how many

Literacy

Oral Language Development
- Use number names

Comprehension
- Make connections using illustrations, prior knowledge, or real-life experiences

Applying the Concept

Distribute the Day 3 activity and crayons. Guide students through the activity by saying:

*Remember that counting by **tens** is a fast way to count. Let's count these fish by tens.*

- *Point to box 1. Let's count the fish. First, let's count how many tens there are. Circle each group of **10** fish as we count: **1, 2, 3, 4**. There are **4** tens. Write the number **4** in the box. It says **4 tens**. Now let's count the fish by tens: **10, 20, 30, 40**. There are **40** fish in all. Circle the number **40**.*

- *Point to box 2. Let's count the fish. First, let's count how many tens there are. Circle each group of **10** fish as we count: **1, 2, 3, 4, 5**. There are **5** tens. Write the number **5** in the box. It says **5 tens**. Now let's count the fish by tens: **10, 20, 30, 40, 50**. There are **50** fish in all. Circle the number **50**.*

- *Point to box 3. Let's count the fish. First, let's count how many tens there are. Circle each group of **10** fish as we count: **1, 2, 3, 4, 5, 6, 7**. There are **7** tens. Write the number **7** in the box. It says **7 tens**. Now let's count the fish by tens: **10, 20, 30, 40, 50, 60, 70**. There are **70** fish in all. Circle the number **70**.*

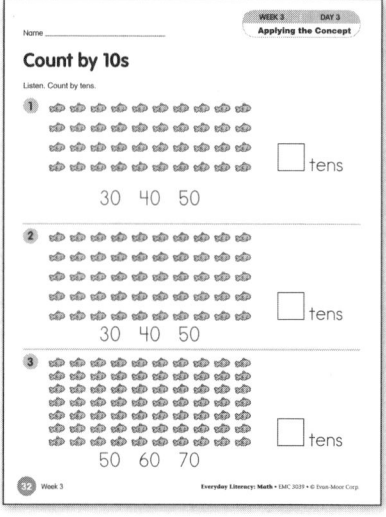

Day 3 activity

Week 3

Everyday Literacy: Math • EMC 3039 • © Evan-Moor Corp.

Day 4
SKILLS

Number Sense
- Count by 2s, 5s, or 10s
- Understand that the final number counted in a set tells how many

Literacy

Oral Language Development
- Respond orally to simple questions
- Use number names

Comprehension
- Make connections using illustrations, prior knowledge, or real-life experiences

Extending the Concept

Introduce the activity by displaying a hundred chart and reviewing. Point to each number as you count:

Let's count by tens to 100: **10, 20, 30, 40, 50, 60, 70, 80, 90, 100.**

Distribute the Day 4 activity and crayons. Guide students through the activity by saying:

- *You are going to count by tens to make a picture. First, put your crayon on the number 10. When you count by tens, what number comes after 10?* (20) *Draw a line from the number 10 to the number 20.*
- *What number comes after 20?* (30) *Yes. Draw a line to the number 30.*
- *What number comes after 30?* (40) *Yes. Draw a line to the number 40.*
- *What number comes after 40?* (50) *Yes. Draw a line to the number 50.*
- *What number comes after 50?* (60) *Yes. Draw a line to the number 60.*
- *What number comes after 60?* (70) *Yes. Draw a line to the number 70.*
- *What number comes after 70?* (80) *Yes. Draw a line to the number 80.*
- *What number comes after 80?* (90) *Yes. Draw a line to the number 90.*
- *What picture did you make?* (a fish) *Yes! Color your fish any color.*

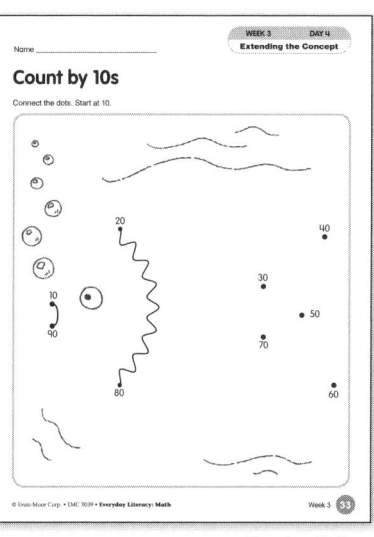

Day 4 activity

Day 5
SKILLS

Number Sense
- Count by 2s, 5s, or 10s
- Understand that the final number counted in a set tells how many

Literacy

Oral Language Development
- Use number names

Mathematical Thinking and Reasoning
- Explore mathematical ideas through song or play

Home–School Connection p. 34
Spanish version available (see p. 2)

Circle Time Math Activity

Reinforce this week's math concept with the following circle time activity:

Have students stand in a circle with their hands clasped together and arms extended, as if about to hit a volleyball.

Stand in the center of the circle and choose the point at which you will start counting by tens.

Face the student who represents your starting point. Tap that student's clasped hands. Go around the circle and continue tapping on subsequent pairs of clasped hands, while singing this song to the tune of "Itsy Bitsy Spider":

10, 20, 30: *Thirty little fingers.*
40, 50, 60: *We can count by tens.*
70, 80, 90: *We are such good singers!*
100 *little fingers. Let's count by tens again!*

Sing again, having a volunteer be the "counter" and starting at a different point in the circle.

© Evan-Moor Corp. • EMC 3039 • **Everyday Literacy: Math**   Week 3

Name _____

WEEK 3 | DAY 1
Confirming Understanding

Count by 10s

Name _____

WEEK 3 | DAY 2
Reinforcing the Concept

Count by 10s

Listen. Count by ones. Then count by tens.

1

ten

☐

2

twenty

☐

3

thirty

☐

© Evan-Moor Corp. • EMC 3039 • *Everyday Literacy: Math* Week 3 **31**

Name _____

WEEK 3 | DAY 3
Applying the Concept

Count by 10s

Listen. Count by tens.

1

☐ tens

30   40   50

2

☐ tens

30   40   50

3

☐ tens

50   60   70

32 Week 3

Everyday Literacy: Math • EMC 3039 • © Evan-Moor Corp.

Name _____

WEEK 3 | DAY 4
Extending the Concept

Count by 10s

Connect the dots. Start at 10.

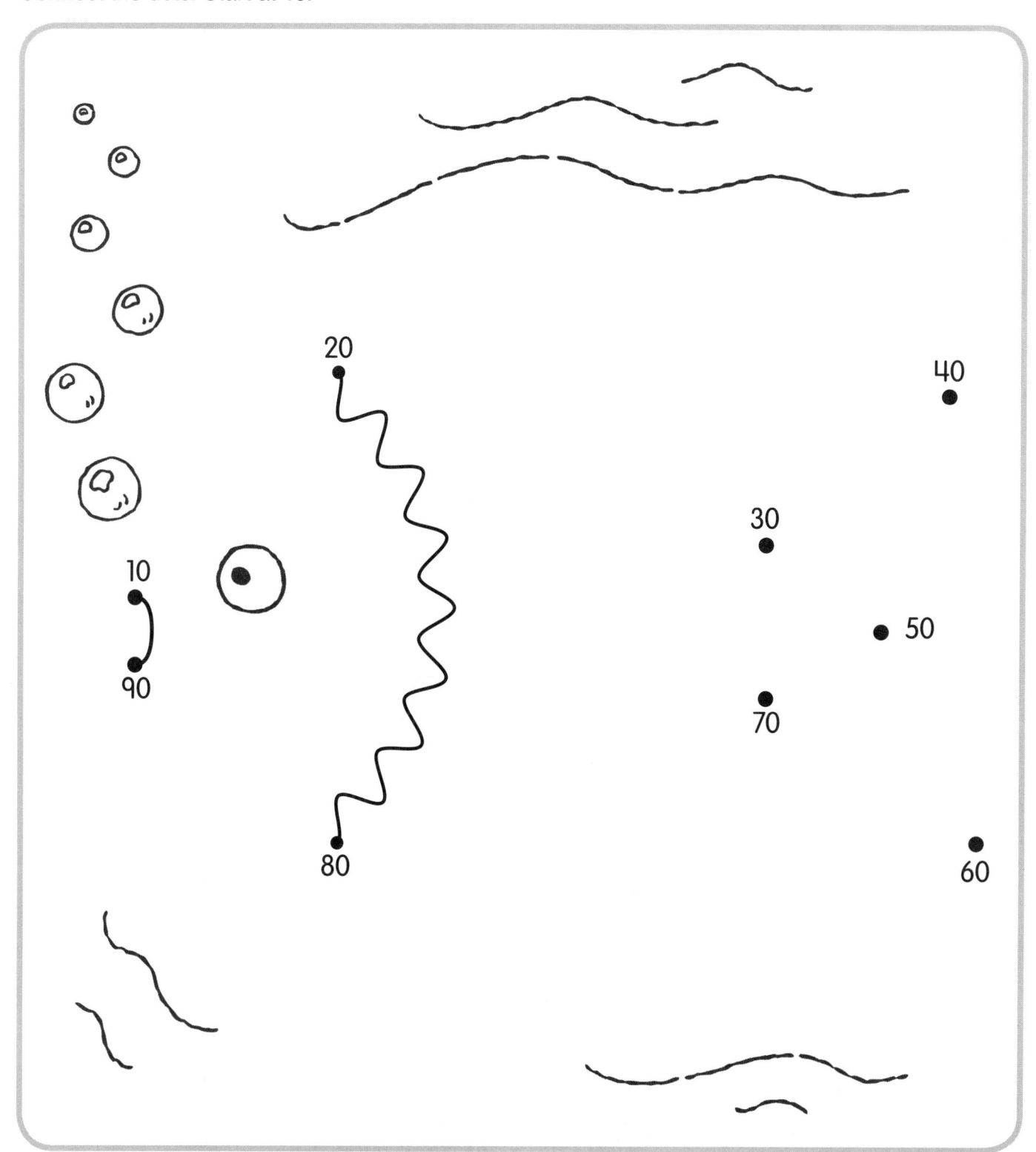

- 20
- 40
- 30
- 50
- 10
- 90
- 70
- 80
- 60

© Evan-Moor Corp. • EMC 3039 • **Everyday Literacy: Math**

Week 3   33

Name _____

What I Learned

WEEK 3

Home–School Connection

What to Do
Have your child look at the picture below. Ask him or her to tell you how many fish are in the fish tanks. Have your child count by tens to find the answer. Ask him or her to write the number of fish. (50)

Math Concept: The number system is based on 10.

To Parents
This week your child learned to count by tens to 100.

What to Do Next
Provide a bowl of 100 beans for your child to count. Have your child make groups of 10 beans. Then have him or her count the beans by tens: *10, 20, 30, 40, 50, 60, 70, 80, 90, 100.*

34 Week 3

Everyday Literacy: Math • EMC 3039 • © Evan-Moor Corp.

WEEK 4

Concept
Skip counting is a strategy for adding and multiplying.

Skip Counting

Math Objective:
To help students skip count by 5s and by 2s

Math Vocabulary:
count by 5s, count by 2s, group, number line, skip count

Day 1 SKILLS

Number Sense
- Use a number line
- Understand that the final number counted in a set tells how many
- Count by 2s, 5s, or 10s

Literacy

Oral Language Development
- Respond orally to simple questions
- Use number names

Comprehension
- Make connections using illustrations, prior knowledge, or real-life experiences
- Answer questions about key details in a text read aloud

Introducing the Concept

To prepare for the lesson, display a pile of 15 loose markers and have 3 rubber bands available. Also display a number line that begins at **0** and ends at **50**. Point to each number on the line as you count it. Say:

*You can use this **number line** to **skip count** by fives. Listen to me count, and watch which numbers I point to: **5, 10, 15, 20, 25, 30, 35, 40, 45, 50**. Now skip count by fives with me.* (students respond)

Call students' attention to the pile of markers. Say:

*This is a big pile of markers. We will count them by fives. But first, we will put them in groups of 5 markers and put a rubber band around each group. Now, as I point to each group, let's count by fives: **5, 10, 15**. How many markers are there in all?* (15) *Counting by fives is faster than counting one at a time.*

Listening to the Story

Distribute the Day 1 activity. Say: *Look at the picture and listen as I read a story about a boy who counts by fives.*

"We have oodles of dry noodles!" said Mr. Ditalini. "You can make fun necklaces." Alex grabbed a string. It had a fat knot at one end. Alex slid **5** elbow macaroni noodles onto the string. Then he put on **5** noodles shaped like wheels. After the wheels, Alex strung **5** noodles that looked like tubes. When he had finished, Alex wondered how many noodles were on his necklace. He counted them in groups of five: **5** tubes, **5** wheels, and **5** elbows: **5, 10, 15**. Alex had **15** noodles on his string!

Confirming Understanding

Distribute crayons. Develop the math concept by asking questions about the story. Ask:

- *What were the first 5 noodles that Alex put on the string?* (elbow macaroni) *Color them.*
- *Why didn't Alex count the noodles one at a time?* (It is faster to count by fives.)
- *Circle each **group** of 5 noodles. How many groups are there?* (3)
- *Now count the noodles by **fives**. How many noodles are there in all?* (15)

Day 1 picture

© Evan-Moor Corp. • EMC 3039 • **Everyday Literacy: Math**

Week 4　35

Day 2 SKILLS

Number Sense
- Use a number line
- Understand that the final number counted in a set tells how many
- Count by 2s, 5s, or 10s

Literacy

Oral Language Development
- Respond orally to simple questions
- Use number names

Comprehension
- Make connections using illustrations, prior knowledge, or real-life experiences

Reinforcing the Concept

Reread the Day 1 story. Then reinforce this week's math concept. Say:

Alex made a necklace. How many noodles were in the pattern that repeated? (5) *How did he count the noodles on his string?* (by fives)

Distribute the Day 2 activity and crayons. Say:

- *Point to box 1. Let's count the noodles. First, let's count them one at a time. Now let's count them five at a time. You know that each row has **5** noodles, so just point to the last noodle of each row: **5, 10, 15**. That was faster than counting by ones! How many noodles are there?* (15) *Circle the number **15**.*

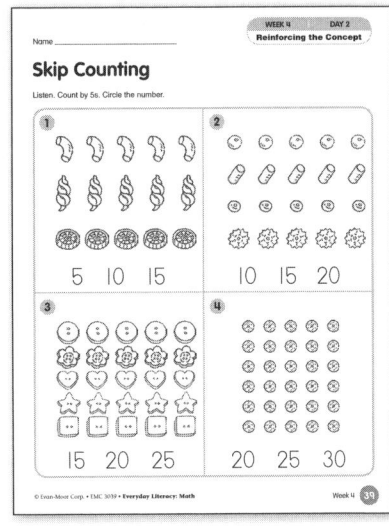

- *Point to box 2. Let's count the beads. This time, let's count by **fives**. Point to each row of beads as you count: **5, 10, 15, 20**. How many beads are there?* (20) *Circle the number **20**.*

- *Point to box 3. Let's count the buttons. Let's count by fives again: **5, 10, 15, 20, 25**. How many buttons are there?* (25) *Circle the number **25**.*

- *Point to box 4. Let's count the sequins. Let's count by fives again: **5, 10, 15, 20, 25, 30**. How many sequins are there?* (30) *Circle the number **30**.*

Day 2 activity

Day 3 SKILLS

Number Sense
- Count by 2s, 5s, or 10s
- Use a number line
- Understand that the final number counted in a set tells how many

Literacy

Oral Language Development
- Respond orally to simple questions
- Use number names

Comprehension
- Make connections using illustrations, prior knowledge, or real-life experiences

Applying the Concept

Display a number line showing the numbers **0** through **50**. Say:

*We know how to count by fives. There's another way to **skip count**. We can count by **twos**.*

Model skip counting by twos, pointing to each number. Distribute the Day 3 activity and crayons. Say:

Point to row 1. Let's count the noodles by twos.

- *Point to the wheels first. Let's count, starting with the number 2: **2, 4, 6, 8, 10**. There are **10** noodles in this row.*

- *Trace each number in this row, starting with the number **2**.*

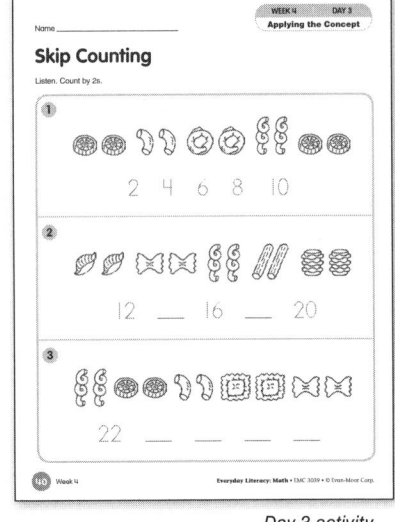

*Point to row 2. Let's keep counting. In row 1, we stopped counting at **10**. When we skip count by twos, what number comes after **10**?* (12) *Yes. Let's count these noodles starting with the number 12: **12, 14, 16, 18, 20**. There are **20** noodles in rows 1 and 2.*

- *Now trace the number **12**. When you skip count by twos, what number comes after **12**?* (14) *Write the number **14** on the line. Trace the number **16**.*

- *When you skip count by twos, what number comes after **16**?* (18) *Write the number **18** on the line. Trace the number **20**.*

Continue the process to count the noodles in row 3.

Day 3 activity

Week 4

Everyday Literacy: Math • EMC 3039 • © Evan-Moor Corp.

Day 4
SKILLS

Number Sense
- Count by 2s, 5s, or 10s

Literacy

Oral Language Development
- Respond orally to simple questions
- Use number names

Comprehension
- Make connections using illustrations, prior knowledge, or real-life experiences

Extending the Concept

Distribute the Day 4 activity and crayons. Guide students through the activity by saying:

- *The mama bird needs to get to her nest. Her two babies are hungry! Help her get to the nest. You will count by* **twos** *to get to the nest.*

- *Put your crayon on the worm. Now draw a line to the number* **2**. *Keep counting by twos. Connect each number with a line. Count along with me:* **2, 4, 6, 8**.

- *Which direction will you go now? Will you go up to the number* **5**? (no) *Why not?* (5 does not come after 8) *Let's keep going straight then, to the number* **10**.

- *Now where will you go? Will you go down to the number* **11**? (no) *Why not?* (We are counting by twos, so the number after 10 is 12.)

- *Keep drawing a line to connect* **10, 12, 14**. *Now which way will you go? Will you go up or down?* (down) *Why?* (16 comes after 14) *Keep counting then:* **16, 18, 20**.

- *Here is the bird's nest! Color the birds and the nest.*

Day 4 activity

Day 5
SKILLS

Number Sense
- Count by 2s, 5s, or 10s

Mathematical Thinking and Reasoning
- Explore mathematical ideas through song or play

Home–School Connection p. 42
Spanish version available (see p. 2)

Circle Time Math Activity

Reinforce this week's math concept with the following circle time activity:

Materials: students' shoes

Activity: Teach students the song below. Then have students sit in a circle with their legs extended. Tell them that you will go around the circle and tap random students on the head (or shoulder). Whoever is tapped should place one shoe in the center of the circle. For the first round, tap 4 students.

Once the shoes are in the center, sing this song to the tune of "Did You Ever See a Lassie?"

Four shoes are in the center, the center, the center.
Four shoes are in the center. So, tell me, please do.
How many toes are now missing their warm shoe?
How many little toes are now missing their shoe?

Remind students that since each shoe can fit 5 toes, they should count each shoe by fives: **5, 10, 15, 20**. Sing: *Twenty little toes are now missing their shoe!*

Play again, changing the number of shoes.

© Evan-Moor Corp. • EMC 3039 • Everyday Literacy: Math

Week 4   37

Name _____

Skip Counting

WEEK 4 | DAY 1
Confirming Understanding

Name _____

WEEK 4 | DAY 2
Reinforcing the Concept

Skip Counting

Listen. Count by 5s. Circle the number.

1

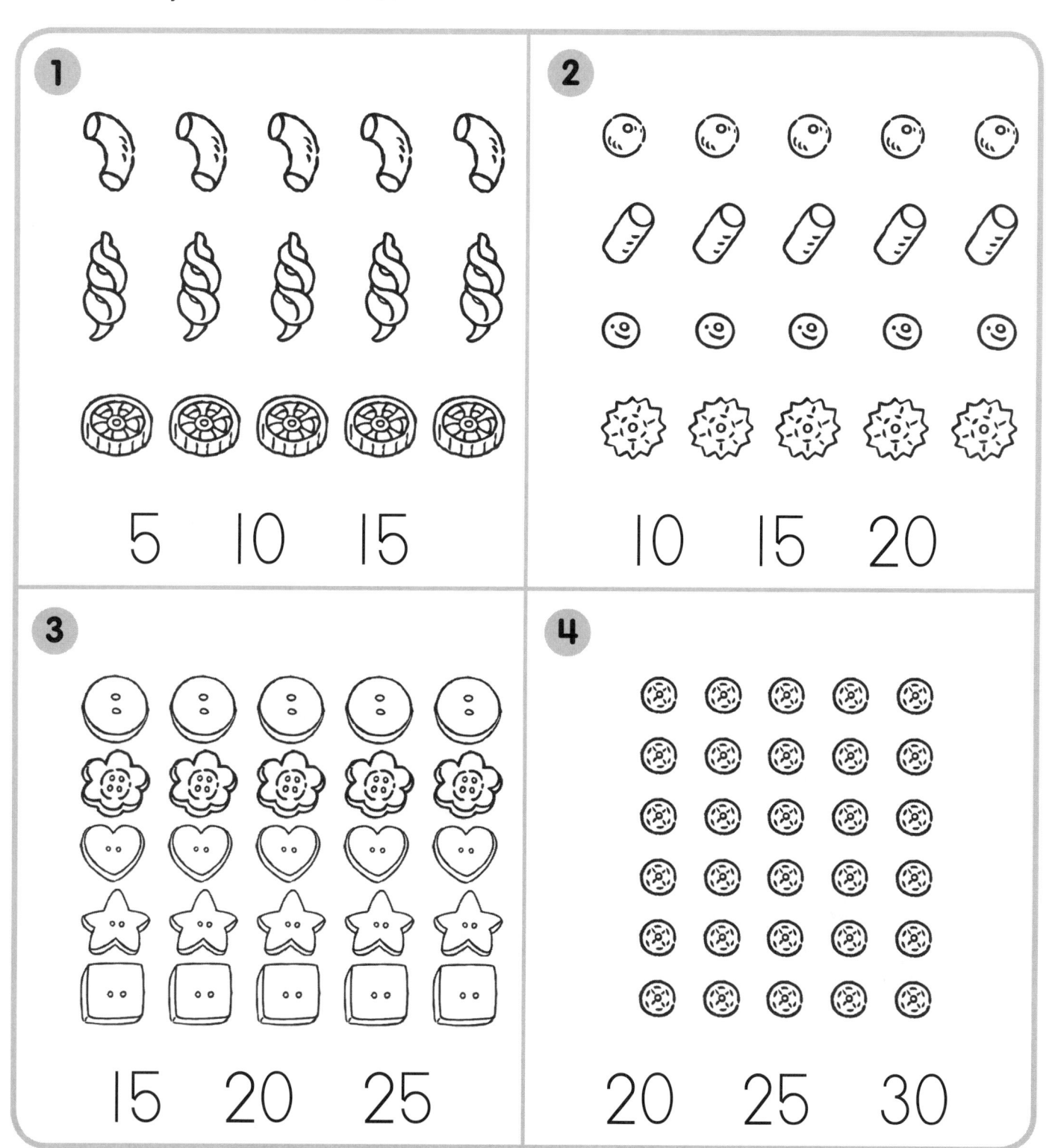

5   10   15

2

10   15   20

3

15   20   25

4

20   25   30

© Evan-Moor Corp. • EMC 3039 • Everyday Literacy: Math

Week 4   39

Name _____

WEEK 4 | DAY 3
Applying the Concept

Skip Counting

Listen. Count by 2s.

1

2 4 6 8 10

2

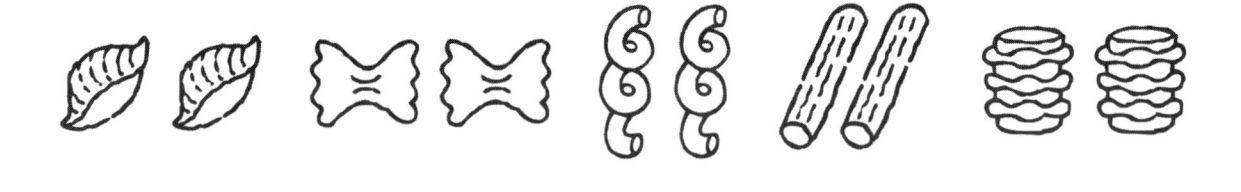

12 ___ 16 ___ 20

3

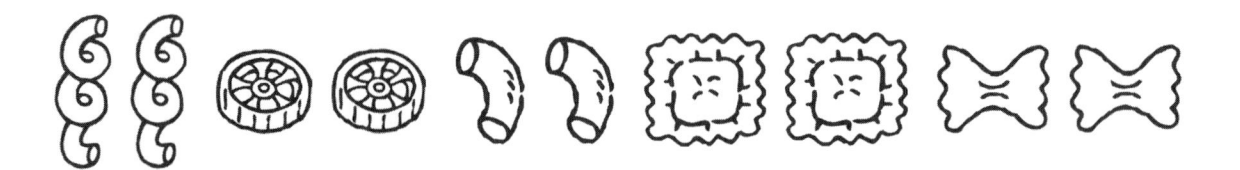

22 ___ ___ ___ ___

40 Week 4

Everyday Literacy: Math • EMC 3039 • © Evan-Moor Corp.

Name _____

WEEK 4 | DAY 4
Extending the Concept

Skip Counting

Listen. Count by 2s.

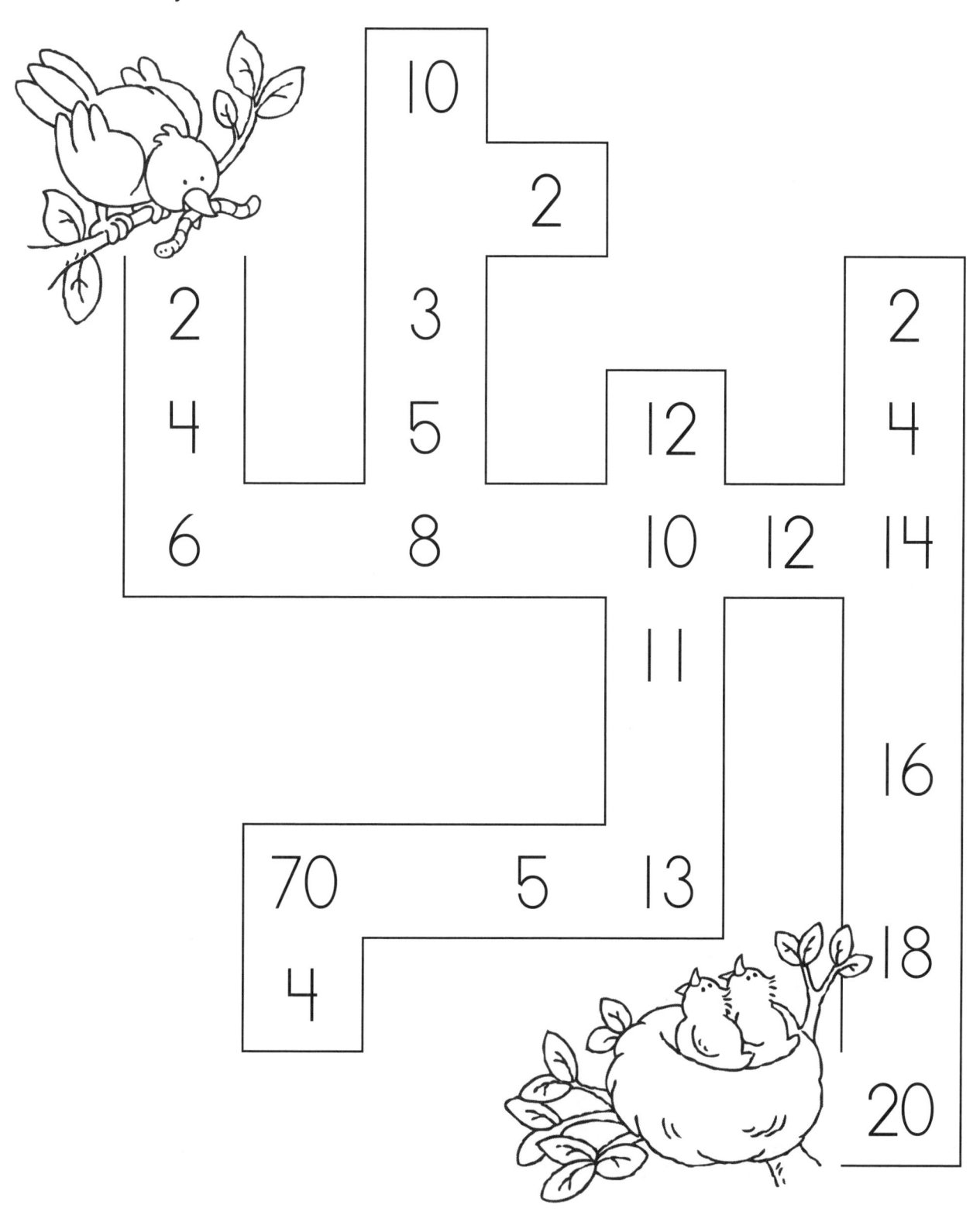

2, 4, 6
3, 5, 8
10, 2, 12
10, 12, 14
11
2, 4
16
70, 5, 13
4
18
20

© Evan-Moor Corp. • EMC 3039 • *Everyday Literacy: Math*   Week 4   41

Name _____

What I Learned

What to Do
Have your child look at the maze below. Ask him or her to count by 2s to get mama bird to her babies' nest. Have your child count aloud and draw a line to connect each number. Then have your child color the mama bird and the babies.

WEEK 4

Home–School Connection

Math Concept: Skip counting is a strategy for adding and multiplying.

To Parents
This week your child learned to count by 5s and 2s.

10
2
2 3 2
4 5 12 4
6 8 10 12 14
11
16
70 5 13 18
4 20

What to Do Next
Ask your child how many fingers and toes you have between the two of you. Stand barefoot to show your toes. Both of you extend your hands. Invite your child to count fingers and toes by fives: 5, 10, 15, 20, 25, 30, 35, 40! Forty fingers and toes!

42 Week 4   Everyday Literacy: Math • EMC 3039 • © Evan-Moor Corp.

WEEK 5

Concept
A 2-digit number represents tens and ones.

Tens and Ones

Math Objective:
To introduce students to the place value of two-digit numbers

Math Vocabulary:
ones, place, tens

Day 1 SKILLS

Number Sense
- Think of whole numbers in terms of tens and ones

Literacy

Oral Language Development
- Use number names
- Respond orally to simple questions

Comprehension
- Make connections using illustrations, prior knowledge, or real-life experiences
- Answer questions about key details in a text read aloud

Introducing the Concept

Before the lesson, gather 12 markers and a marker box that holds 10 markers. Also, draw two columns on the board. Label the left column **tens** and the right column **ones**. Point to the markers and say:

I have 12 markers. I can fit 10 of them in the box. I have one group of 10 markers. How many more markers are there? (2)

Write a **1** in the tens column. Then say:

I can show these markers as a number. The 1 is in the tens place to show that there is one group of 10 markers. I will write 2 in the ones column. The 2 is in the ones place to show how many more markers there are. How many markers do I have in all? (12)

Listening to the Story

Distribute the Day 1 activity. Say: *Look at the picture and listen as I read a story about a boy putting balls into groups.*

Each student is helping to clean up after P.E. Manny is taking care of the balls. He puts the balls into a stretchy bag. He counts the balls as he places them in the bag. **Ten** balls fit inside. **Five** balls are left over. Manny begins to write down the number of balls. He thinks, "I have one group of **10** balls so I'll write a **1**. Then I'll write a **5** because there are **5** more balls. We have **15** balls in all."

Confirming Understanding

Distribute crayons. Develop the math concept by asking questions about the story. Ask:

- *How many groups of 10 balls does Manny make?* (1)
- *Look at the number on Manny's paper. How many tens are there?* (1) *How many ones?* (5) *What number does that make?* (15)
- *Draw 2 more balls. In the boxes, write the number that tells how many balls there are now.* (17)

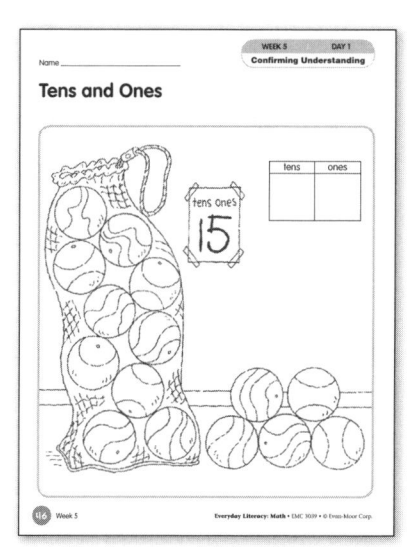

Day 1 picture

© Evan-Moor Corp. • EMC 3039 • Everyday Literacy: Math

Week 5   43

Day 2
SKILLS

Number Sense
- Think of whole numbers in terms of tens and ones

Literacy

Oral Language Development
- Respond orally to simple questions
- Count by 2s, 5s, or 10s

Comprehension
- Make connections using illustrations, prior knowledge, or real-life experiences
- Answer questions about key details in a text read aloud

Reinforcing the Concept

Reread the Day 1 story. Then reinforce this week's math concept by discussing the story. Say:

We learned to put items into groups of 10. How many groups of 10 balls did Manny have? (1) Let's put more items in groups of 10.

Distribute the Day 2 activity and crayons. Say:

- *Point to number 1. Let's count how many groups of 10 there are. Each row has 10 crayons. How many groups of 10 are there? (2) Write the number 2 in the first box. Let's read: 2 tens. How many crayons are there in all? Count by tens: 10, 20. Write the number 20 in the next box.*

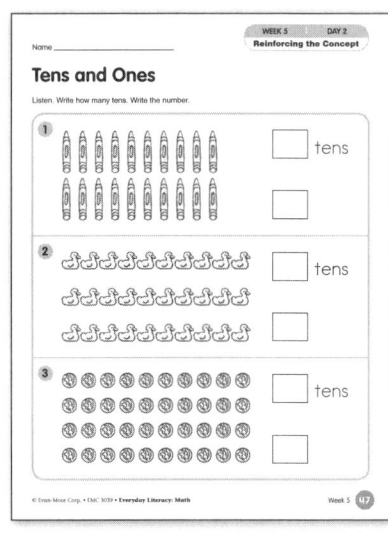

- *Point to number 2. Let's count how many groups of 10 there are. Each row has 10 rubber duckies. How many groups of 10 are there? (3) Write the number 3 in the first box. Let's read: 3 tens. How many duckies are there in all? Count by tens: 10, 20, 30. Write the number 30 in the next box.*

- *Point to number 3. Let's count how many groups of 10 there are. Each row has 10 marbles. How many groups of 10 are there? (4) Write the number 4 in the first box. Let's read: 4 tens. How many marbles are there in all? Count by tens: 10, 20, 30, 40. Write the number 40 in the next box.*

Day 2 activity

Day 3
SKILLS

Number Sense
- Think of whole numbers in terms of tens and ones

Literacy

Oral Language Development
- Respond orally to simple questions

Comprehension
- Make connections using illustrations, prior knowledge, or real-life experiences

Applying the Concept

Introduce the activity by reminding students of what they have learned so far. Say:

We know how to group items into tens. Now we will group items into tens and ones.

Distribute the Day 3 activity and crayons. Say:

Point to number 1. Each tall column, or line, has 10 stars. Count how many tens there are. (2) Write the number 2 in the tens box. How many more stars are there? (3) Write the number 3 in the ones box.

- *What is the number? (23) That's right. Twenty-three is made of 2 tens and 3 ones.*

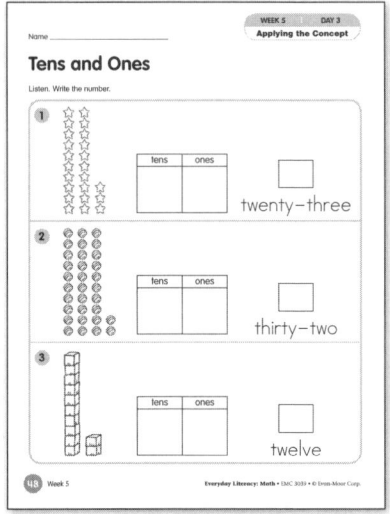

- *Write the number 23 in the box labeled with the word twenty-three.*

Repeat the process with the 32 balls and the 12 blocks.

Day 3 activity

44 Week 5   Everyday Literacy: Math • EMC 3039 • © Evan-Moor Corp.

Day 4 SKILLS

Number Sense
- Think of whole numbers in terms of tens and ones

Literacy

Oral Language Development
- Respond orally to simple questions

Comprehension
- Make connections using illustrations, prior knowledge, or real-life experiences

Extending the Concept

Distribute the Day 4 activity and crayons. Guide students through the activity by saying:

*Point to number 1. It shows candy. Some of the candy is in bags. Each bag has **10** candies. Some of the candy is loose. These candies are the **ones**.*

- *How many **tens** are there? (4) How many **ones** are there? (2)*
- *Show the candies as a number. Point to the box next to the candies. It has the words **tens** and **ones** at the top. Since we have 4 tens, write the number **4** in the **tens** column. We have 2 more candies, so write the number **2** in the **ones** column. What is the number? (42) That's right. **Forty-two** is made of **4 tens** and **2 ones**. Write the number 42 in the smaller box.*

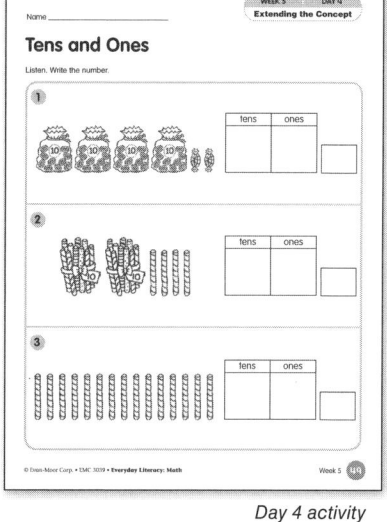

Repeat the process with the candy sticks in number 2. Then guide students through number 3. Say:

*Point to number 3. It shows candy sticks, but this time they are not in groups of 10. Make your own groups of 10. Count **10** candies. Draw a circle around that group of 10. How many **tens** did you circle? (1) Write **1** in the **tens** column. How many candies are left over? (5) Write **5** in the **ones** column. What is the number? (15) Write 15 in the box.*

Day 4 activity

Day 5 SKILLS

Number Sense
- Think of whole numbers in terms of tens and ones

Mathematical Thinking and Reasoning
- Select and use various types of reasoning and methods of proof

Home–School Connection p. 50
Spanish version available (see p. 2)

Hands-on Math Activity

Reinforce this week's math concept with the following hands-on activity:

Materials: empty egg cartons (1 for each small group), marker, lima beans, paper

Preparation: Label each egg carton cell with the numbers **1** through **9**. (Three numbers will be repeated.)

Activity: Divide students into small groups. Give each group an egg carton and 2 lima beans. Explain that they will play a game called "Scrambled Numbers." Give each student a sheet of paper. Have them fold their papers in half lengthwise to make 2 columns. Label the left column **tens** and the right column **ones**.

Explain the rules of the game: Place the 2 lima beans in the egg carton. Close the lid. Shake the carton. Open the lid. Note the numbers that the lima beans landed on. If the lima beans landed on a **4** and a **3**, for example, the number could be **34** or **43**. The player who shook the carton decides which number everyone will write. The team members should write that number on their papers in the correct columns.

Have players take turns scrambling the numbers.

© Evan-Moor Corp. • EMC 3039 • **Everyday Literacy: Math** Week 5

Name _____

WEEK 5 | DAY 1
Confirming Understanding

Tens and Ones

tens	ones

tens ones
15

46 Week 5   Everyday Literacy: Math • EMC 3039 • © Evan-Moor Corp.

Name _____

WEEK 5 | DAY 2
Reinforcing the Concept

Tens and Ones

Listen. Write how many tens. Write the number.

1. ☐ tens

   ☐

2. ☐ tens

   ☐

3. ☐ tens

   ☐

© Evan-Moor Corp. • EMC 3039 • *Everyday Literacy: Math*

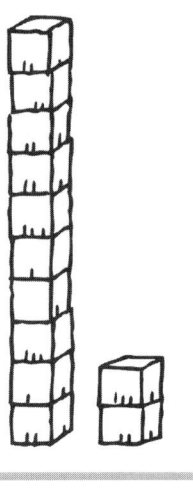

Name _____

WEEK 5 | DAY 3
Applying the Concept

Tens and Ones

Listen. Write the number.

1. twenty-three

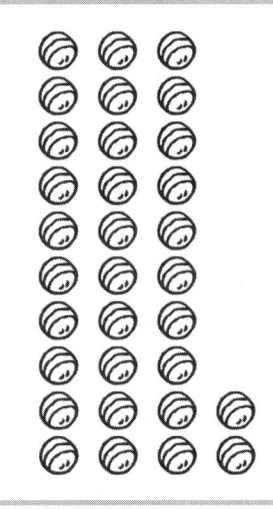

2. thirty-two

3. twelve

48 Week 5

Everyday Literacy: Math • EMC 3039 • © Evan-Moor Corp.

Tens and Ones

Listen. Write the number.

1

tens	ones

2

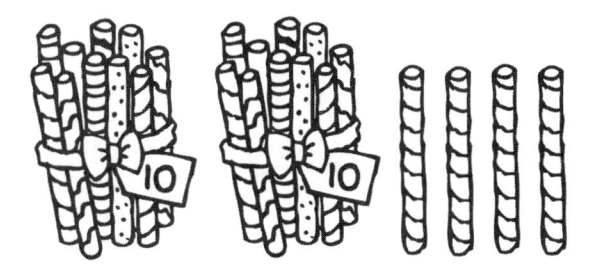

tens	ones

3

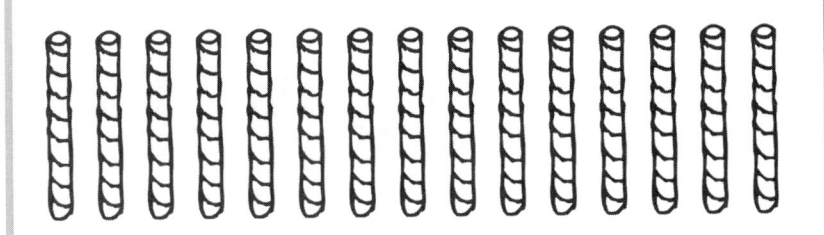

tens	ones

WEEK 5 | DAY 4
Extending the Concept

Week 5
49

Name _____

What I Learned

WEEK 5

Home–School Connection

Math Concept: A 2-digit number represents tens and ones.

To Parents
This week your child learned the place value of 2-digit numbers.

What to Do
Have your child look at the picture below and tell you how many balls are in the bag. (10) Then have him or her tell you how many balls are outside the bag. Have your child write the number of balls in the tens and ones box. Ask: *How many balls are there in all?* (15)

tens	ones

What to Do Next
Provide a bowl of up to 99 lima beans for your child to group into tens and ones. Have him or her use a tens and ones box, like the one above, to write the number. Provide a different number of beans each time until your child has written four numbers in the boxes.

50 Week 5

Everyday Literacy: Math • EMC 3039 • © Evan-Moor Corp.

WEEK 6

Concept
Addition is the operation of *adding to*.

How Many In All?

Math Objective:
To help students understand the meaning of the operation of addition

Math Vocabulary:
add, addition sentence, equals, in all, plus, problem, sum

Day 1 SKILLS

Algebra
- Use more than one strategy to solve a problem
- Determine the unknown whole number in an equation

Literacy

Oral Language Development
- Respond orally to simple questions
- Use mathematical terms

Comprehension
- Make connections using illustrations, prior knowledge, or real-life experiences
- Answer questions about key details in a text read aloud

Introducing the Concept

Explain to students that they are going to **add**, or put together, two numbers to get a larger number. Present this simple word problem:

Corky has 3 bones and King has 2 bones. How many bones do the dogs have in all? One way to find out is to draw bones and count.

Draw 3 bones and 2 bones. Have students count to find the answer. Then write the addition sentence for the problem: **3 + 2 = 5**. Say:

I used numbers from the problem to write an addition sentence. This sentence shows us that 3 plus 2 equals 5. Read the sentence with me: 3 plus 2 equals 5. How many bones do the dogs have in all? (5) *That's right. The answer, or the* **sum***, is 5.*

Present a simple addition problem. Have pairs of students draw circles to find the sum. Then have them write the addition sentence and read it out loud.

Listening to the Story

Distribute the Day 1 activity. Say: *Look at the picture and listen as I read a story about a boy named Charlie who drew circles as a way to add.*

---

*Gramps asked Charlie about his baseball games. "I got **2** hits on Saturday, and I got **4** hits on Sunday," said Charlie. "Wow! That's a lot of hits!" said Gramps. Charlie boasted, "I can tell you exactly how many hits I made." He drew **2** circles for the hits he made on Saturday. Then he drew **4** circles for the hits he made on Sunday. Charlie counted the circles and said, "See, **2** hits plus **4** hits equals **6** hits!"*

Confirming Understanding

Distribute pencils. Develop the math concept by asking questions about the story:

- *What does Charlie want to figure out?* (how many hits he made altogether) *What does he do to find the answer?* (draws circles and counts)
- *In the space on Charlie's paper, write an addition sentence that tells about Charlie's hits.* (2 + 4 = 6)
- *Let's read our addition sentence out loud: 2 plus 4 equals 6.*

Day 1 picture

Day 2 SKILLS

Algebra
- Use more than one strategy to solve a problem
- Determine the unknown whole number in an equation

Literacy

Oral Language Development
- Respond orally to simple questions
- Use mathematical terms

Comprehension
- Make connections using illustrations, prior knowledge, or real-life experiences
- Answer questions about key details in a text read aloud

Reinforcing the Concept

Reread the Day 1 story. Then reinforce this week's math concept by discussing the story. Say:

Our story was about a boy who told a baseball story. What addition sentence did he use to tell his story? (2 + 4 = 6)

Distribute the Day 2 activity and pencils. Ask students to listen to this story problem:

Charlie had 1 baseball. He found 4 more. How many did he have in all?

- *Write an addition sentence to tell this story. First, trace the 1. This is the number of baseballs Charlie started with.*
- *In the box, trace the plus sign to show that we will add.*
- *Trace the 4. This is how many more baseballs he found.*
- *Trace the equal sign.*
- *Finally, trace the 5. This is the answer. The answer in an addition sentence is called the sum. How many baseballs are there in all?* (5)

Repeat the steps above, narrating simple addition stories about the bats, the hot dogs, and the pretzels. Have students trace or write the missing elements of the addition sentences.

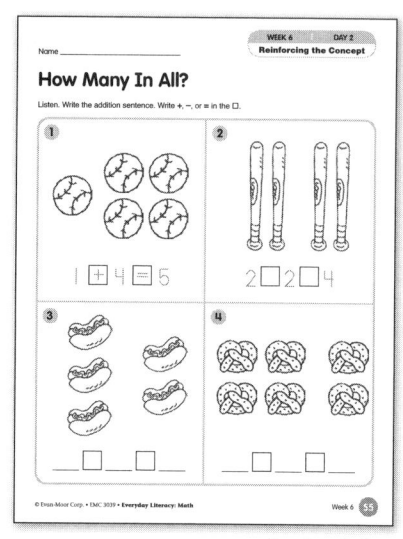

Day 2 activity

Day 3 SKILLS

Algebra
- Use more than one strategy to solve a problem
- Determine the unknown whole number in an equation

Literacy

Oral Language Development
- Use mathematical terms

Comprehension
- Make connections using illustrations, prior knowledge, or real-life experiences

Applying the Concept

Distribute the Day 3 activity and pencils. Guide students through the activity by saying:

- *Let's solve the addition sentence about popsicles at the top of the page. 4 + 1 = ____. Write the sum in the box.* (5) *Look at the addition sentence on the other side of the popsicles. It looks a little different, but it says the same thing: 4 + 1 = ____. Write the sum in the box.* (5)
- *Point to number 1. It shows 2 popsicles plus 4 popsicles. Write the sum.* (6) *Now look at the addition sentences in the other column. Find the addition sentence that says the same thing: 2 + 4 = ____. Write the sum.* (6) *Draw a line to match the two addition sentences that tell the same story.*
- *Point to number 2. It shows 3 popsicles plus 4 popsicles. Write the sum.* (7) *Now go to the other column. Find the addition sentence that says the same thing: 3 + 4 = ____. Write the sum.* (7) *Draw a line to match the two addition sentences that tell the same story.*
- *Point to number 3. It shows 3 popsicles plus 2 popsicles. Write the sum.* (5) *Now go to the other column. Find the addition sentence that says the same thing: 3 + 2 = ____. Write the sum.* (5) *Draw a line to match the two addition sentences that tell the same story.*

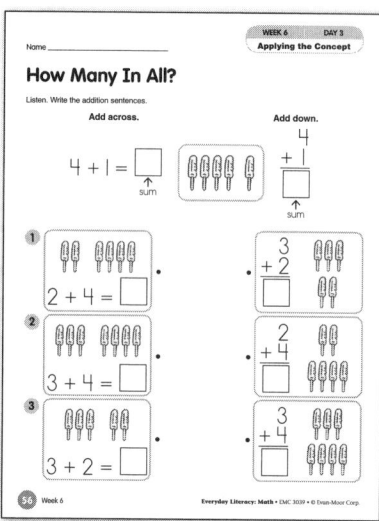

Day 3 activity

52 Week 6

Everyday Literacy: Math • EMC 3039 • © Evan-Moor Corp.

Day 4 SKILLS

Algebra
- Use more than one strategy to solve a problem
- Determine the unknown whole number in an equation

Literacy

Oral Language Development
- Respond orally to simple questions
- Use mathematical terms

Comprehension
- Make connections using illustrations, prior knowledge, or real-life experiences

Extending the Concept

Distribute the Day 4 activity and pencils. Guide students through the activity by saying:

- *Look at the picture at the top of the page. It tells a story.* **Chase and Ryan were playing ball. Four more friends came to play. How many kids are there in all?**

- *Let's solve this problem. First, we need to figure out whether we* **put together**, *or add, or* **take away**, *or subtract. Are we putting together friends or taking away friends?* (putting together) *That means we will* **add**.

- *Read the sentences:* **I will add. I will subtract.** *Underline the correct sentence.* (I will add.)

- *Now we are going to write the addition sentence two ways. First, let's write the sentence going* **across**. *In the first box, write the number that tells how many children were there to start.* (2) *Then in the circle, write the symbol that means* **add** *(plus sign). In the next box, write how many friends came to join the first two* (4). *In the next circle, write an equal sign. In the last box, write the* **sum** (6). *Read the sentence:* **2 + 4 = 6**.

- *Now let's write the same addition sentence going* **down**. *Instead of an equal sign, there's a line.* (students write) *Read the sentence:* **2 + 4 = 6**.

Day 4 activity

Day 5 SKILLS

Algebra
- Use more than one strategy to solve a problem

Literacy

Oral Language Development
- Use mathematical terms

Mathematical Thinking and Reasoning
- Explore mathematical ideas through song or play

Home–School Connection p. 58
Spanish version available (see p. 2)

Hands-on Math Activity

Reinforce this week's math concept with the following hands-on activity:

Materials: plastic bat, plastic baseball, glove, and any other baseball-related props

Preparation: Take students to a large,w open area. Then teach them the song below, to the tune of "Take Me Out to the Ball Game."

Activity: Tell students they will model an addition sentence by acting out the words of the song:

*Let's go out and play baseball.
Let's put together a team.
Start with a batter who swings a bat.
That's 1 player. We need more than that!
So let's add 8 more to the ball team.
Now how many are we?
We are 9 great players. We'll win. You just come and see!*

You can have different students playing the part of the batter, and then add different quantities of players, which will result in a different sum. Alternatively, you may ask students to create a drawing and label it with an addition sentence (1 + 8 = 9) to represent the story in the song.

© Evan-Moor Corp. • EMC 3039 • *Everyday Literacy: Math*   Week 6   53

Name _____

WEEK 6 | DAY 1
Confirming Understanding

How Many In All?

Name _____

WEEK 6 | DAY 2
Reinforcing the Concept

How Many In All?

Listen. Write the addition sentence. Write +, −, or = in the ☐.

1

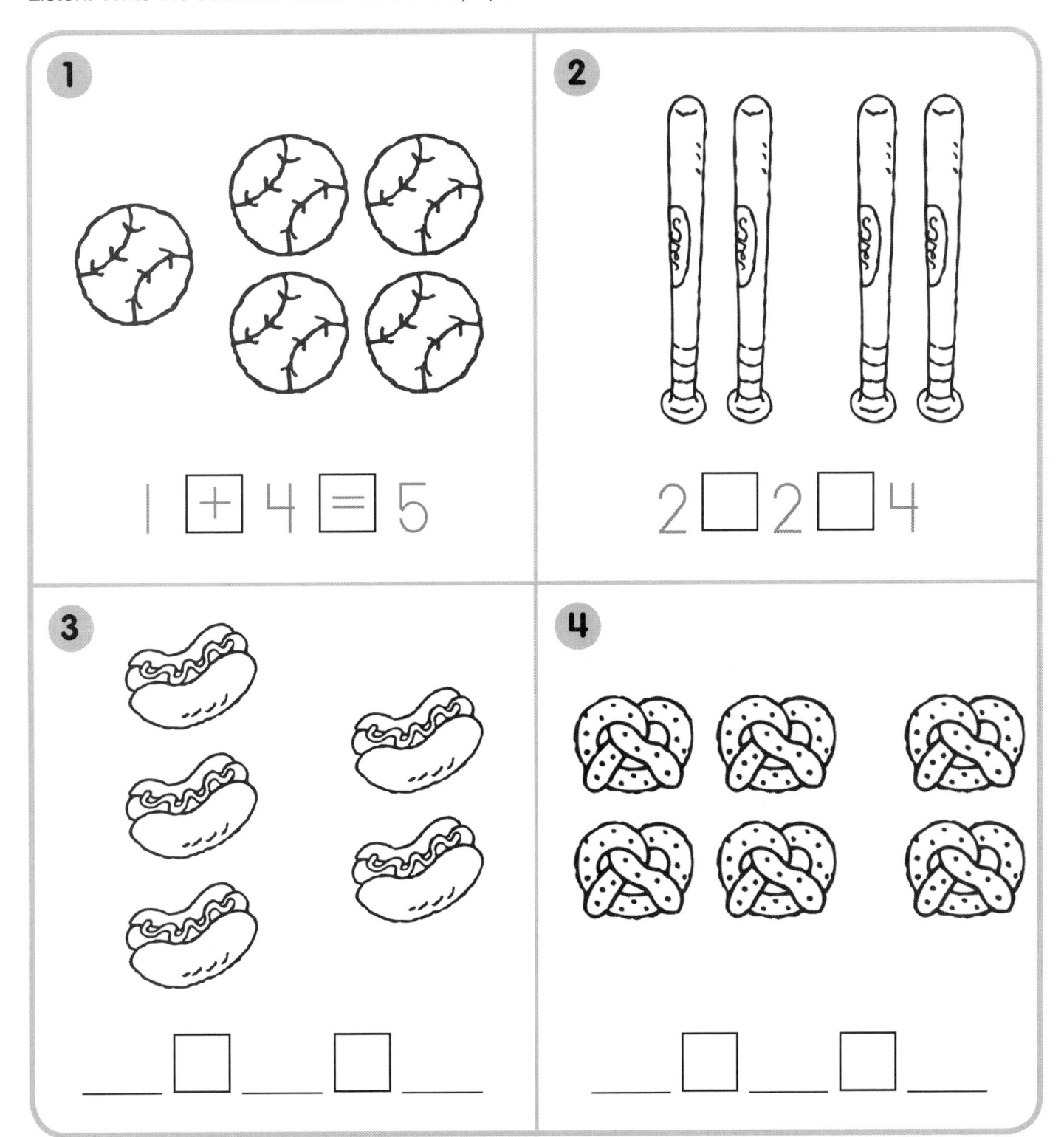

1 ☐+☐ 4 ☐=☐ 5

2

2 ☐ 2 ☐ 4

3

___ ☐ ___ ☐ ___

4

___ ☐ ___ ☐ ___

© Evan-Moor Corp. • EMC 3039 • **Everyday Literacy: Math**   Week 6   55

Name _____

WEEK 6 | DAY 3
Applying the Concept

How Many In All?

Listen. Write the addition sentences.

Add across.

$4 + 1 = \square$
↑ sum

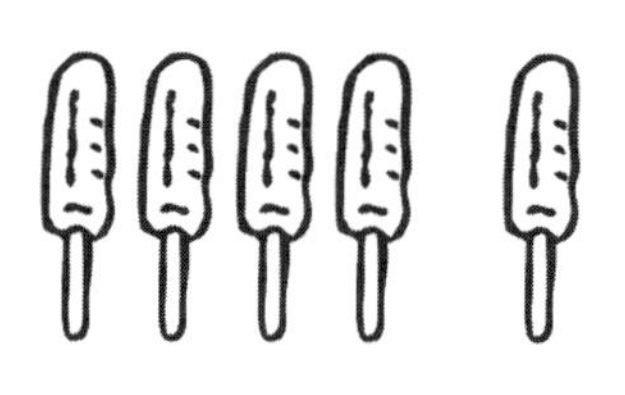

Add down.

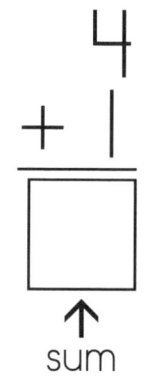

$$\begin{array}{r} 4 \\ +\ 1 \\ \hline \square \end{array}$$
↑ sum

1 $2 + 4 = \square$

$$\begin{array}{r} 3 \\ +\ 2 \\ \hline \square \end{array}$$

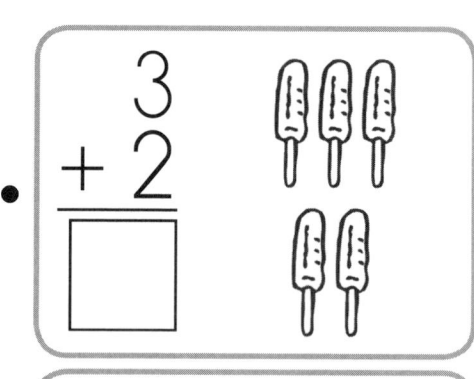

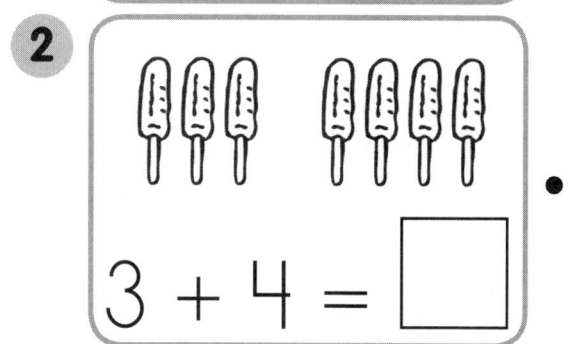

2 $3 + 4 = \square$

$$\begin{array}{r} 2 \\ +\ 4 \\ \hline \square \end{array}$$

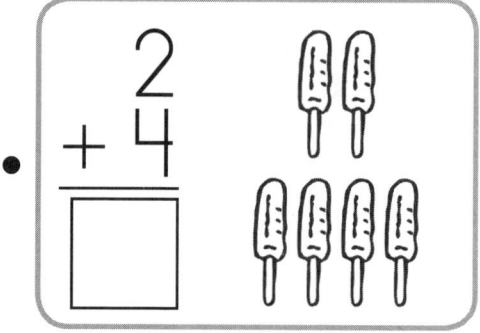

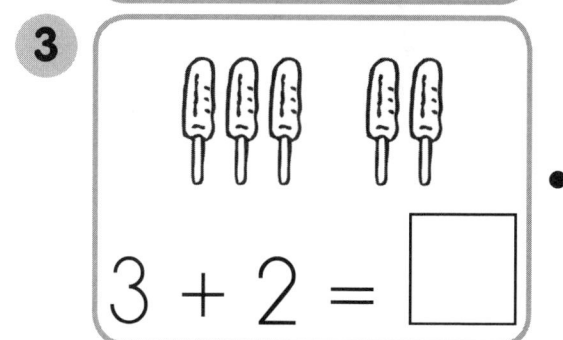

3 $3 + 2 = \square$

$$\begin{array}{r} 3 \\ +\ 4 \\ \hline \square \end{array}$$

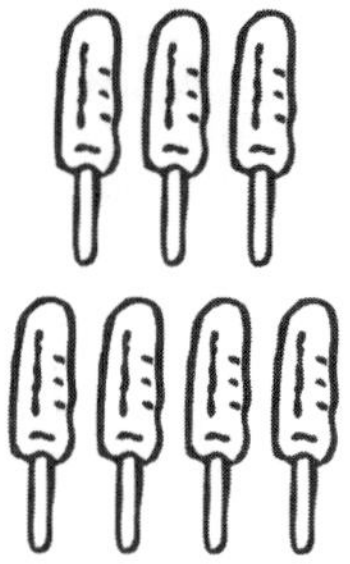

Name _____

WEEK 6 | DAY 4
Extending the Concept

How Many In All?

Listen. Underline.

I will add.

I will subtract.

Add across.

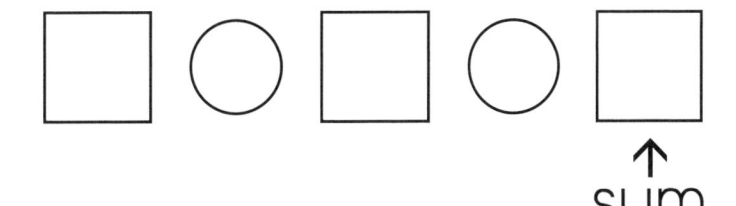

↑ sum

Add down.

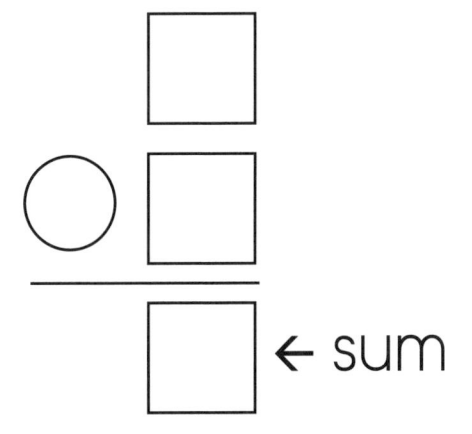

← sum

Week 6

Name _____

What I Learned

What to Do
Have your child look at the addition sentences below. Ask him or her to solve each problem and write the sum. Then have your child draw lines from the horizontal addition sentences to the vertical addition sentences to match the sums.

WEEK 6

Home–School Connection

Math Concept: Addition is the operation of *adding to*.

To Parents
This week your child learned to use different strategies to add.

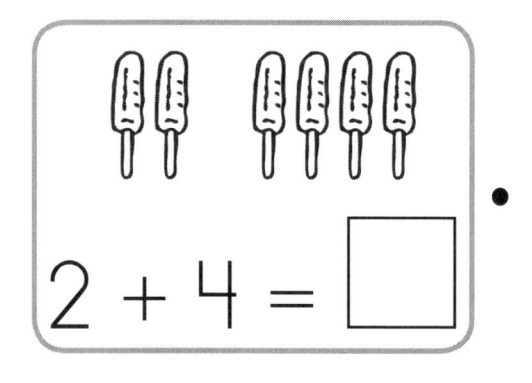

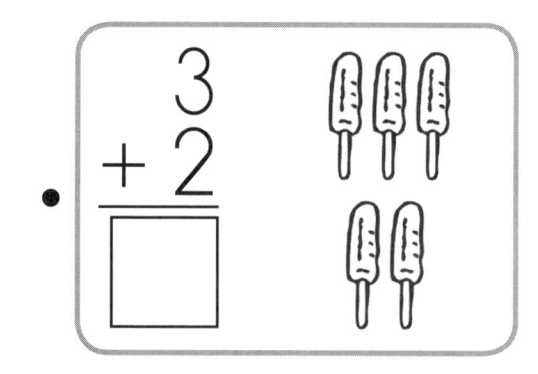

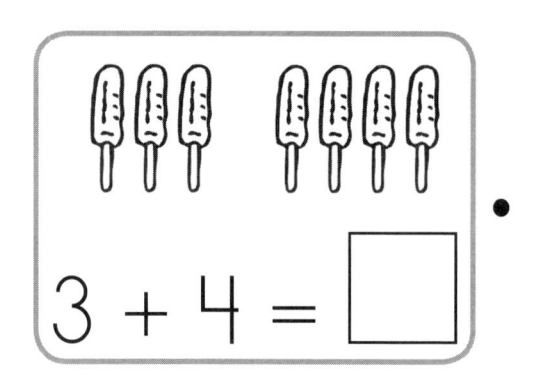

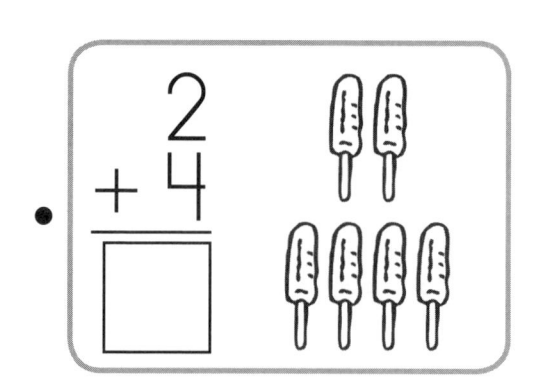

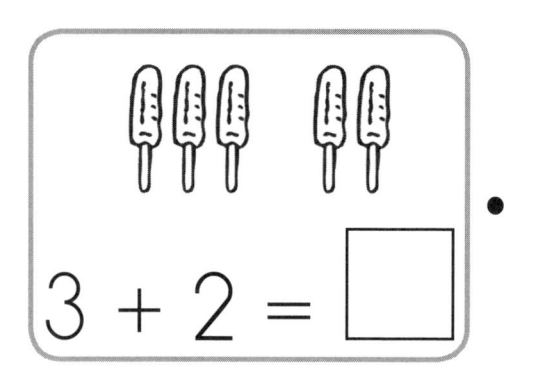

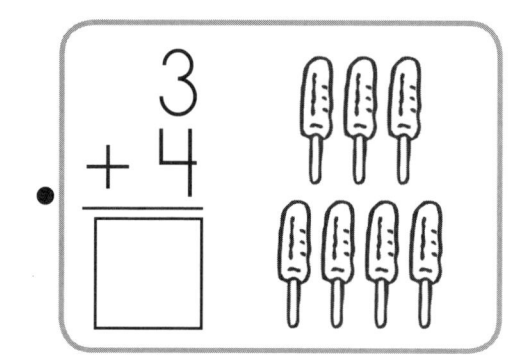

2 + 4 = ☐

3 + 4 = ☐

3 + 2 = ☐

3 + 2 = ☐

2 + 4 = ☐

3 + 4 = ☐

What to Do Next
Display 10 cotton balls. Divide the cotton balls into 2 groups. Have your child count how many are in each group and write an addition sentence that tells the story of the cotton balls.

WEEK 7

Concept
Subtraction is the operation of *taking away*.

How Many Are Left?

Math Objective:
To help students understand the operation of subtraction

Math Vocabulary:
difference, equals, left, minus, subtract, subtraction sentence

Day 1 SKILLS

Algebra
- Use more than one strategy to solve a problem
- Determine the unknown whole number in an equation

Literacy

Oral Language Development
- Respond orally to simple questions
- Use mathematical terms

Comprehension
- Make connections using illustrations, prior knowledge, or real-life experiences
- Answer questions about key details in a text read aloud

Introducing the Concept

Tell students they are going to **subtract**, or take away, a smaller number from a larger number to see how many are **left**. Give a simple word problem. Say:

I have 7 crackers. I give 3 away. How many crackers do I have left? One way to find out is to draw crackers and cross out.

Draw 7 crackers on the board and cross out 3. Have students count the remainder. Write the equation that stands for the problem: 7 − 3 = 4. Say:

*I used the numbers in the problem to write a subtraction sentence. This subtraction sentence shows that **7 minus 3 equals 4**. Read the sentence with me: **7 minus 3 equals 4**. How many crackers do I have left?* (4)

Have students repeat the steps above to solve a simple subtraction problem. Let them work in pairs and then read their subtraction sentence out loud.

Listening to the Story

Distribute the Day 1 activity. Say: *Look at the picture and listen as I read a story about the way a girl solves a subtraction problem.*

Karla's grandma is coming to visit. Grandma has to drive **9** miles to get to Karla's house. Grandma called and said she had driven **4** miles. She asked Karla to figure out how many more miles she had left to drive. Karla made a drawing. She drew one line for each mile. Then Karla crossed out **4** lines because Grandma had already driven **4** miles. Next, she counted the lines left over. "You're **5** miles away, Grandma! See you soon!" said Karla.

Confirming Understanding

Distribute pencils. Develop the math concept by asking questions about the story:

- *What does Karla want to find out?* (how many miles Grandma has left to drive)
- *Karla drew a line for each mile. Cross out **4** lines on Karla's paper to show the miles Grandma has already driven.*
- *On the line at the bottom of Karla's paper, write the subtraction sentence for the problem.* (9 − 4 = 5)
- *Read the subtraction sentence out loud:* **9 minus 4 equals 5**.

Day 1 picture

© Evan-Moor Corp. • EMC 3039 • **Everyday Literacy: Math**

Week 7

Day 2 SKILLS

Algebra
- Use more than one strategy to solve a problem
- Determine the unknown whole number in an equation

Literacy

Oral Language Development
- Respond orally to simple questions
- Use mathematical terms

Comprehension
- Make connections using illustrations, prior knowledge, or real-life experiences
- Answer questions about key details in a text read aloud

Reinforcing the Concept

Reread the Day 1 story. Then reinforce this week's math concept by discussing the story. Say:

Karla found out how many miles Grandma had left to drive. Did she add or subtract? (subtract)

Distribute the Day 2 activity and pencils. Say:

Point to box 1. **Grandma baked 5 cupcakes. Karla ate 2. How many cupcakes are left?** *Show the story with a subtraction sentence.*

- *First, trace the* **5**. *This is the number of cupcakes Grandma baked.*
- *Write a minus sign in the box to show that we will subtract.*
- *Trace the number* **2**. *This is how many cupcakes Karla ate.*
- *Write an equal sign in the next box.*
- *Finally, trace the number* **3**. *This is the answer. The answer in a subtraction problem is called the* **difference**. *How many cupcakes are left?* (3)

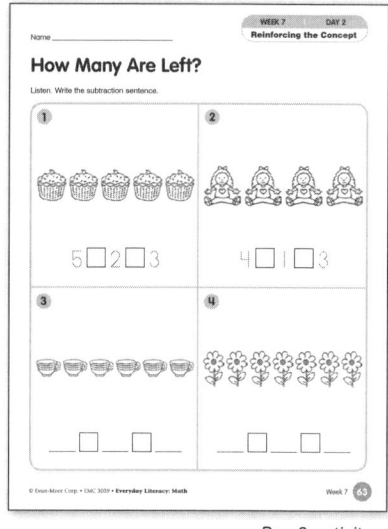

Repeat the steps above, narrating simple subtraction stories about the dolls, the teacups, and the flowers, and having students trace or write the missing elements of the subtraction sentences.

Day 2 activity

Day 3 SKILLS

Algebra
- Use more than one strategy to solve a problem
- Determine the unknown whole number in an equation

Literacy

Oral Language Development
- Respond orally to simple questions
- Use mathematical terms

Comprehension
- Make connections using illustrations, prior knowledge, or real-life experiences

Applying the Concept

Introduce the concept of subtracting with zero. Model with blocks or markers, and say:

When we subtract, we take away. When we subtract **all** *the items from a group, we are left with nothing, or* **zero**. *When we subtract* **zero** *from a group, we are left with the same number we started with.*

Distribute the Day 3 activity and pencils. Guide students through the activity by saying:

- *Point to number 1. How many butterflies are in the first group?* (5) *Cross out* **all** *5 butterflies. How many did you take away?* (5) *Now how many are* **left**? (0)
- *Trace the subtraction sentence below the first group of butterflies. Then in the box, write the answer, or the* **difference**: 0.
- *Point to the second group of butterflies. How many butterflies are there?* (5) *How many are taken away?* (none, or zero) *How many are* **left**? (5) *Trace the subtraction sentence below the second group of butterflies. In the box, write the answer, or the* **difference**: 5.

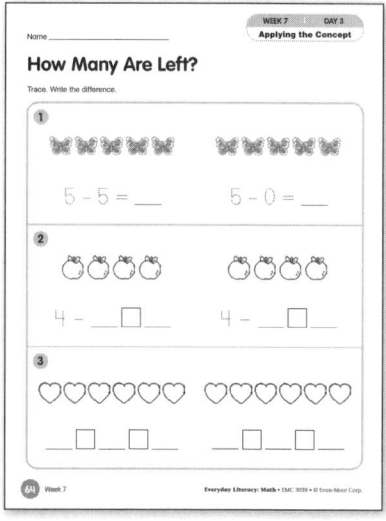

Repeat these steps for the apples and hearts. Have students subtract **all** items in the first group and subtract **zero** items from the second group. Have them trace or write the missing elements of the subtraction sentences.

Day 3 activity

60 Week 7   Everyday Literacy: Math • EMC 3039 • © Evan-Moor Corp.

Day 4 SKILLS

Algebra
- Use more than one strategy to solve a problem
- Determine the unknown whole number in an equation

Literacy

Oral Language Development
- Respond orally to simple questions
- Use mathematical terms

Comprehension
- Make connections using illustrations, prior knowledge, or real-life experiences

Extending the Concept

Distribute the Day 4 activity and pencils. Then narrate this subtraction story:

I had 6 chocolate chip cookies. My sister ate 2. How many cookies are left?

- *This story is shown in the first subtraction sentence: 6 – 2 = ___. In this sentence, you subtract across. Write the **difference** in the box.* (4)
- *You can also show this story with a picture. There are 6 cookies and 2 are crossed out.*
- *Look at the other subtraction sentence. It looks a little different because you subtract down instead of across. But it says the same thing: 6 – 2 = ___. Write the **difference** in the box.* (4)

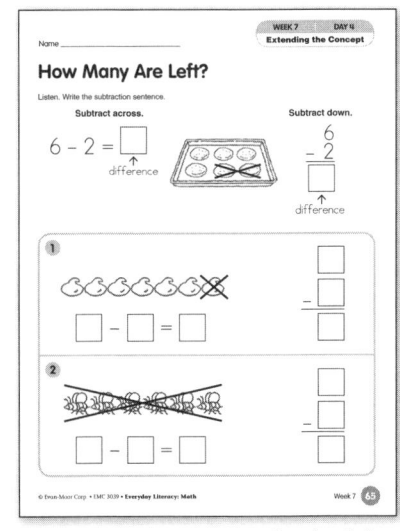

Point to number 1. It tells another story: **My cookie had 7 chocolate chips. I took 1 chocolate chip off the cookie. How many chocolate chips are still in the cookie?**

- *Tell the story two ways. First, subtract across. In the boxes below the chips, write 7 – 1 = 6. Next, tell the story by subtracting down. Write 7 – 1 = 6.*

Repeat the process with the bees, narrating a simple word problem and having students subtract across and down to show 6 – 6 = 0.

Day 4 activity

Day 5 SKILLS

Algebra
- Use more than one strategy to solve a problem

Literacy

Oral Language Development
- Use mathematical terms

Mathematical Thinking and Reasoning
- Select and use various types of reasoning and methods of proof

Home–School Connection p. 66
Spanish version available (see p. 2)

Hands-on Math Activity

Reinforce this week's math concept with the following hands-on activity:

Materials: small index cards, pencils, crayons, hole punch

Preparation: Divide students into groups of three, with each student assuming the role of Student A, B, or C.

Activity: Give an index card to each student. Explain that each member of the group will represent a word problem in a different way. Student A will illustrate the problem with pictures. Student B will show the problem with a subtraction sentence written horizontally. Student C will show the problem with a subtraction sentence written vertically.

Read the following word problem to students: **Six bees were buzzing over some flowers. Three flew away. How many bees are left?**

After students have finished, display the sets of pictures and subtraction sentences, either by making mobiles or by mounting each set of three index cards on construction paper.

Name _____

WEEK 7 | DAY 1
Confirming Understanding

How Many Are Left?

Name _____

WEEK 7 | DAY 2
Reinforcing the Concept

How Many Are Left?

Listen. Write the subtraction sentence.

1

5 ☐ 2 ☐ 3

2

4 ☐ 1 ☐ 3

3

___ ☐ ___ ☐ ___

4

___ ☐ ___ ☐ ___

© Evan-Moor Corp. • EMC 3039 • *Everyday Literacy: Math*   Week 7   63

Name _____

WEEK 7 | DAY 3
Applying the Concept

How Many Are Left?

Trace. Write the difference.

1

5 − 5 = ___   5 − 0 = ___

2

4 − ☐ ___   4 − ☐ ___

3

___ ☐ ___ ☐ ___   ___ ☐ ___ ☐ ___

64 Week 7   Everyday Literacy: Math • EMC 3039 • © Evan-Moor Corp.

Name _____

WEEK 7 | DAY 4
Extending the Concept

How Many Are Left?

Listen. Write the subtraction sentence.

Subtract across.

6 − 2 = ☐
↑
difference

Subtract down.

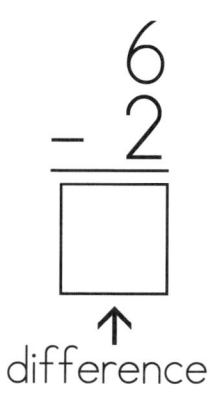

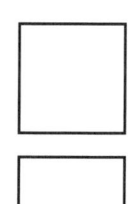

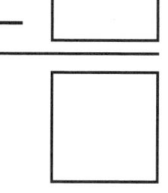

6
− 2
___
☐
↑
difference

1

☐ − ☐ = ☐

☐
− ☐
___
☐

2

☐ − ☐ = ☐

☐
− ☐
___
☐

© Evan-Moor Corp. • EMC 3039 • *Everyday Literacy: Math* Week 7 65

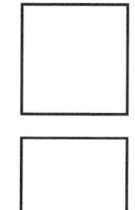

Name _____

What I Learned

What to Do
Have your child look at the pictures in each box and tell you a story about each picture. For example, he or she might say, *We bought 5 cupcakes and then we ate 2. How many cupcakes are left?* Have your child write a subtraction sentence for each picture, encouraging him or her to make up sentences for numbers 3 and 4.

WEEK 7

Home–School Connection

Math Concept: Subtraction is the operation of *taking away*.

To Parents
This week your child learned several ways to subtract.

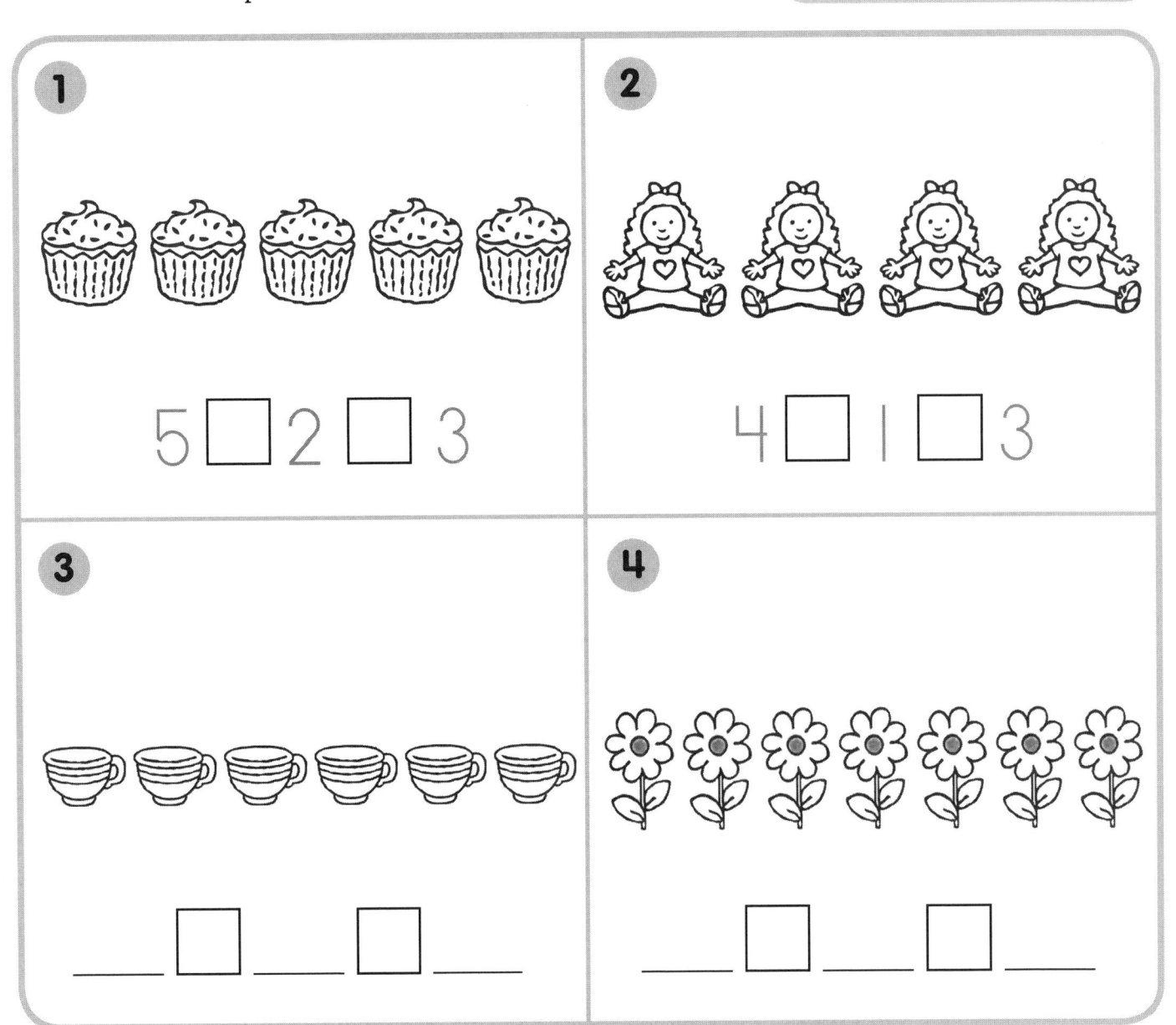

1

5 ☐ 2 ☐ 3

2

4 ☐ 1 ☐ 3

3

___ ☐ ___ ☐ ___

4

___ ☐ ___ ☐ ___

What to Do Next
Help your child find pictures that show sets of 10 or fewer, such as those found in a weekly supermarket mailing. Have him or her cross out random numbers and then make up a word problem, such as, *There were 6 bottles of jam on the shelf. We bought 2. How many are left?*

WEEK 8

Concept
Mathematical operations have properties.

Order of Addends

Math Objective:
To help students understand and apply properties of addition

Math Vocabulary:
addend, addition sentence, order, zero

Day 1 SKILLS

Algebra
- Understand the commutative property of addition
- Solve addition problems

Literacy

Oral Language Development
- Respond orally to simple questions
- Use mathematical terms

Comprehension
- Make connections using illustrations, prior knowledge, or real-life experiences
- Answer questions about key details in a text read aloud

Introducing the Concept

Have 2 boys and 5 girls line up in a row in the front of the room. Then ask:

How many boys are in line? (2) *How many girls?* (5) *How many students in all?* (7) *What* **addition sentence** *tells about these students?* (2 + 5 = 7)

Write the sentence on the board. Indicate that the **2** and the **5** are called **addends**. Then change the order of students so the girls are first. Repeat the process above and end by writing the addition sentence 5 + 2 = 7. Then ask:

How are the addition sentences the same? (same addends and sum)
How are they different? (the order of the addends) *So **2 + 5 = 7** and **5 + 2 = 7**. You can add numbers in **any order**. The answer is the same.*

Repeat the process with a new combination of different students.

Listening to the Story

Distribute the Day 1 activity. Before reading the story, show students a few dominoes. Then say: *Look at the picture and listen as I read a story about a boy who uses a game to practice addition.*

Joey used dominoes to practice adding. His first domino had **4** dots on the top and **5** dots on the bottom. Mom asked Joey to add the dots together. Joey thought, and then he said, "**Four** plus **5** equals **9**." Joey picked another domino. It had **5** dots on the top and **4** dots on the bottom. Joey saw that this domino had the same number of dots as the first domino. This time Joey did not have to think. "**Nine**," he said. "The number of dots is the same, so the answer is the same!"

Confirming Understanding

Distribute pencils. Develop the math concept by asking questions about the story. Ask:

- *What is the same about the two dominoes?* (number of dots) *What is different?* (order of the dots)

- *What is the sum of the dots on the first domino?* (9) *What is the sum of the dots on the second domino?* (9)

- *Why do the dots on both dominoes add up to 9?* (There are the same number of dots.)

- *Write an addition sentence for each domino in the box below it.* (4 + 5 = 9; 5 + 4 = 9)

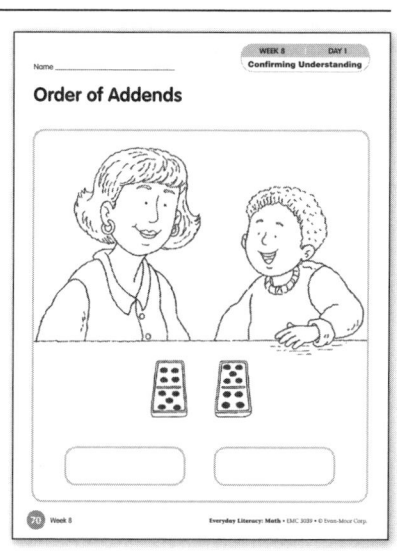

Day 1 picture

Day 2 SKILLS

Algebra
- Understand the commutative property of addition
- Solve addition problems

Literacy

Oral Language Development
- Respond orally to simple questions
- Use mathematical terms

Comprehension
- Make connections using illustrations, prior knowledge, or real-life experiences
- Answer questions about key details in a text read aloud

Reinforcing the Concept

Reread the Day 1 story. Then reinforce this week's math concept by discussing the story. Say:

Our story told about the order of numbers in an addition sentence. What are these numbers called? (addends)

Distribute the Day 2 activity and pencils. Say:

- *Point to the first domino. It shows two groups of dots: **6** dots and **2** dots. How many dots are there altogether?* (8) *Yes, the sum is **8**. Write the number **8** in the box after the equal sign.*

- *Now find the addition sentence in the next column that has the same **addends** in a different order. What does the sentence say?* (2 + 6) *Draw a line to the addition sentence that says **2 + 6**.*

- *How many dots does that domino have altogether?* (8) *Write the number **8** in the box next to the equal sign. So **6 + 2** and **2 + 6** have the same sum: **8**.*

Repeat the process with addition sentences 2 and 3.

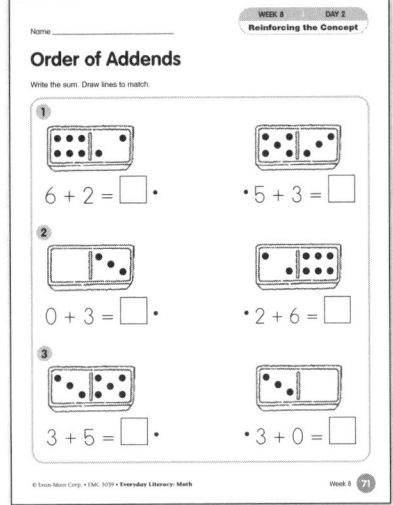

Day 2 activity

Day 3 SKILLS

Algebra
- Understand the commutative property of addition
- Solve addition problems

Literacy

Oral Language Development
- Respond orally to simple questions
- Use mathematical terms

Comprehension
- Make connections using illustrations, prior knowledge, or real-life experiences

Applying the Concept

Distribute the Day 3 activity and pencils. Then introduce the activity by reviewing:

- *Point to number 1. There are two groups of toy soldiers: **2** and **3**. How many toy soldiers are there altogether?* (5) *Yes, **2 + 3 = 5**. In the box after the equal sign, write the sum: **5**.*

- *Look at number 2. There are **3** soldiers and **2** soldiers. Do you think they will equal the same number of soldiers as problem number 1?* (yes) *Why?* (The addends are the same, just in a different order.)

- *Write the addition sentence **3 + 2 = 5**.*

- *Is the sum the same for addition sentences number 1 and number 2?* (yes) *Yes, **2 + 3** and **3 + 2** have the same sum: **5**.*

- *Now point to number 3. There are **4** checkers and **1** checker. How many checkers are there in all?* (5) *That's right. The addition sentence that tells this is **4 + 1 = 5**. Below the line, write the sum: **5**.*

- *Look at number 4. There is **1** checker and **4** checkers. How many checkers are there in all?* (5) *That's right. Write the addition sentence that tells this: **1 + 4 = 5**. Notice that **4 + 1** and **1 + 4** have the same sum: **5**.*

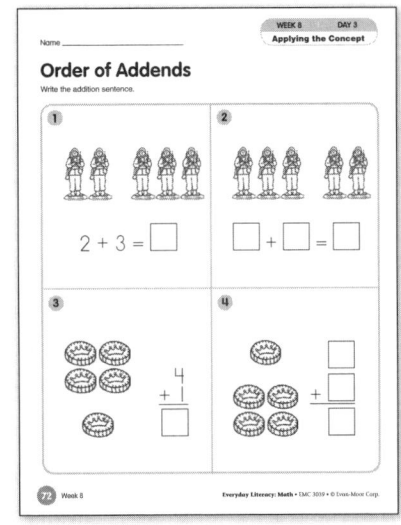

Day 3 activity

Week 8

Everyday Literacy: Math • EMC 3039 • © Evan-Moor Corp.

Day 4 SKILLS

Algebra
- Understand the commutative property of addition
- Solve addition problems

Literacy

Oral Language Development
- Respond orally to simple questions
- Use mathematical terms

Comprehension
- Make connections using illustrations, prior knowledge, or real-life experiences

Extending the Concept

Distribute the Day 4 activity and pencils. Then introduce the activity by saying:

- *Look at number 1. It shows **4** balls and **2** balls. How many balls are there in all? (6) Write an addition sentence to tell this story. (4 + 2 = 6)*

- *In the box below, draw **2** balls and **4** balls. How many balls are there in all? (6)*

- *Below your drawing, write the addition sentence **2 + 4 = 6**.*

- *Look at the addition sentences: **4 + 2** and **2 + 4** have the same sum: **6**.*

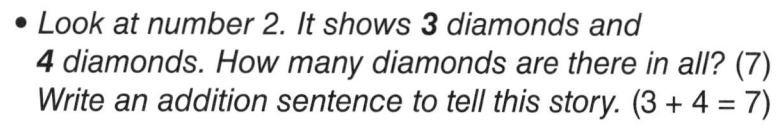

- *Look at number 2. It shows **3** diamonds and **4** diamonds. How many diamonds are there in all? (7) Write an addition sentence to tell this story. (3 + 4 = 7)*

- *In the box below, draw **4** diamonds and **3** diamonds. How many diamonds are there in all? (7)*

- *Below your drawing, write the addition sentence **4 + 3 = 7**. Notice that **3 + 4** and **4 + 3** have the same sum: **7**.*

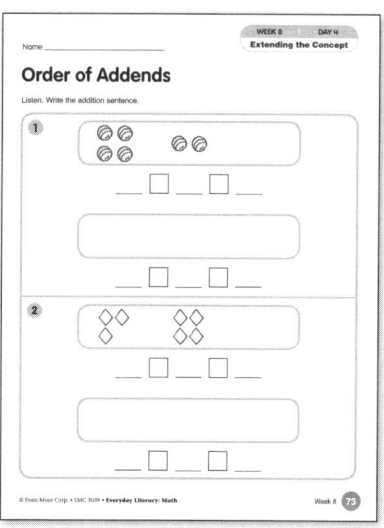

Day 4 activity

Day 5 SKILLS

Algebra
- Understand the commutative property of addition
- Solve addition problems

Mathematical Thinking and Reasoning
- Explore mathematical ideas through song or play

Home–School Connection p. 74
Spanish version available (see p. 2)

Hands-on Math Activity

Reinforce this week's math concept with the following hands-on activity:

Materials: index cards

Preparation: Prepare sets of cards for groups to play "Addition Concentration." Write an addition fact, such as **2 + 3**, on each index card. For each addition fact, make a card with the related addition fact: **3 + 2**. Make a set of **16** cards for each small group.

Activity: Divide students into small groups. Give each group a deck of the cards. Have them shuffle the cards and place them facedown in four rows of four cards.

Have students take turns drawing two cards that have the same addends in different order. For example, if the player draws **5 + 3**, he or she should state the addition fact and give the sum: **5 + 3 = 8**. He or she should draw another card. If this card is **3 + 5**, there is a match and the player keeps both cards.

If the player draws a different card, for example, **2 + 4**, he or she should state the addition fact and the sum: **2 + 4 = 6**. Since there is no match, the cards go back facedown in the same spot. The player with the most cards wins.

© Evan-Moor Corp. • EMC 3039 • *Everyday Literacy: Math* Week 8 69

Name _____

WEEK 8 | DAY 1
Confirming Understanding

Order of Addends

Week 8

Name _____

WEEK 8 | DAY 2
Reinforcing the Concept

Order of Addends

Write the sum. Draw lines to match.

1. 
6 + 2 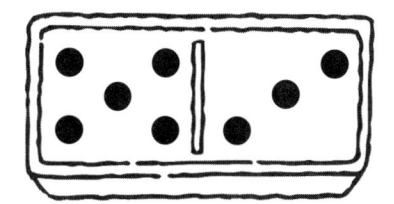 = ☐ •   • 5 + 3 = ☐

2. 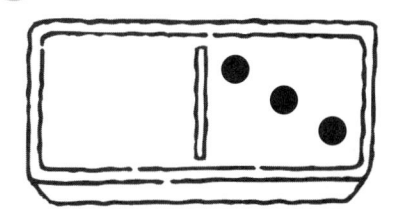
0 + 3 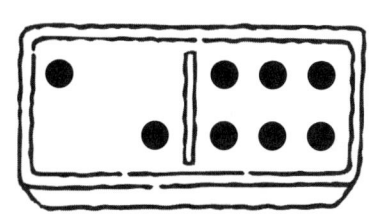 = ☐ •   • 2 + 6 = ☐

3. 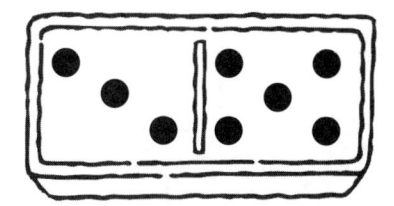
3 + 5 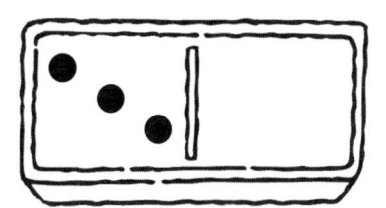 = ☐ •   • 3 + 0 = ☐

© Evan-Moor Corp. • EMC 3039 • Everyday Literacy: Math

Week 8 71

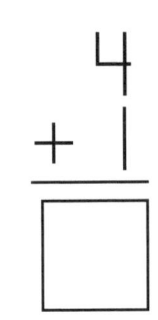

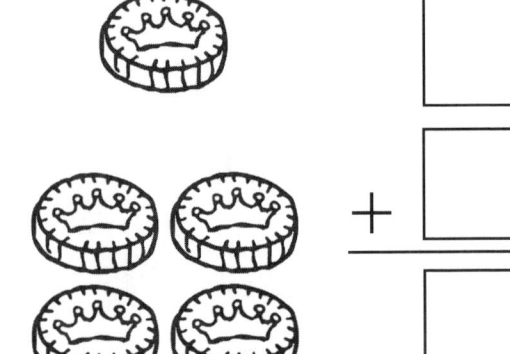

Order of Addends

Write the addition sentence.

1

2 + 3 = ☐

2

☐ + ☐ = ☐

3

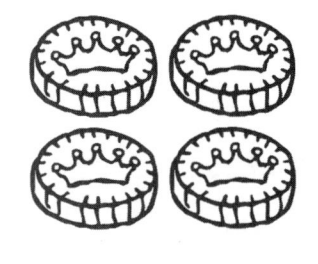

```
  4
+ 1
———
  ☐
```

4

```
  ☐
  ☐
+ ———
  ☐
```

WEEK 8 | DAY 3
Applying the Concept

Name _____

WEEK 8 | DAY 4
Extending the Concept

Order of Addends

Listen. Write the addition sentence.

1

___ ☐ ___ ☐ ___

___ ☐ ___ ☐ ___

2

___ ☐ ___ ☐ ___

___ ☐ ___ ☐ ___

Week 8

Name _____

What I Learned

WEEK 8

Home–School Connection

What to Do
Have your child look at the picture below and tell you about the number of the dots on the dominoes. Then have him or her write the two addition sentences that tell about the dots.
(4 + 5 = 9 and 5 + 4 = 9)

Math Concept: Mathematical operations have properties.

To Parents
This week your child learned that the order of addends does not affect the sum.

What to Do Next
Write the numbers 0 through 5 on 6 lima beans. Place them in a paper bag. Have your child shake the bag and take out 2 beans. Have your child say two different addition sentences using the two numbers, such as 2 + 5 = 7 and 5 + 2 = 7.

WEEK 9

Concept
Addition and subtraction have an inverse relationship.

Fact Families

Math Objective:
To help students understand the relationship between numbers in a fact family

Math Vocabulary:
addend, fact family, parts, related, solve, whole

Day 1 SKILLS

Algebra
- Understand the relationship between addition and subtraction

Literacy

Oral Language Development
- Respond orally to simple questions
- Use mathematical terms

Comprehension
- Make connections using illustrations, prior knowledge, or real-life experiences
- Answer questions about key details in a text read aloud

Introducing the Concept

Display these related facts: 3 + 6 = 9, 6 + 3 = 9; 9 – 3 = 6, 9 – 6 = 3.
Use pictures to review the addition and subtraction operations. Then ask:

- *What do you notice about these four number sentences?* (They have the same three numbers. They make addition and subtraction sentences.)
- *The numbers 3, 6, and 9 belong to a **fact family**. These three numbers are **related**, just as the members of a family are related. **Three** and **6** are the parts. **Nine** is the whole. There are other fact families, like **2**, **3**, and **5**.*

Listening to the Story

Distribute the Day 1 activity. Say: *Look at the picture and listen as I read a poem about the numbers **2**, **3**, and **5**: a fact family that lives in the same house.*

*There once was a family of facts.
Three numbers to add or subtract.
They all were related.
Never separated.
Four sentences made just like that!*

Confirming Understanding

Distribute crayons. Develop the math concept by asking questions about the poem. Say:

- *Look at the bushes by the fact family house. What symbol means **add**?* (plus sign) *Color the bush with a plus sign green. Trace the addition sentence below the fact family house.* (2 + 3 = 5)
- *Now write an addition sentence that uses the same numbers as the first one, but put the addends in a different order.* (3 + 2 = 5)
- *What symbol means **subtract**?* (minus sign) *Color the bush with the minus sign blue. Trace the subtraction sentence below the fact family house.* (5 – 3 = 2)
- *Write the last subtraction sentence that uses the fact family. It starts with the biggest number, just like the other subtraction sentence. Write the sentence.* (5 – 2 = 3)

Day 1 picture

© Evan-Moor Corp. • EMC 3039 • **Everyday Literacy: Math** Week 9 75

Day 2
SKILLS

Algebra
- Understand the relationship between addition and subtraction

Literacy

Oral Language Development
- Respond orally to simple questions
- Use mathematical terms

Comprehension
- Make connections using illustrations, prior knowledge, or real-life experiences
- Answer questions about key details in a text read aloud

Reinforcing the Concept

Reread the Day 1 poem. Then reinforce this week's math concept by discussing the poem. Say:

*Our poem was about the numbers **2**, **3**, and **5**, which make a **fact family**. If you know **2 + 3 = 5**, what other addition fact do you know? (3 + 2 = 5) What is one subtraction fact that is related to this addition fact? (5 – 3 = 2 or 5 – 2 = 3)*

Distribute the Day 2 activity and pencils. Say:

- *Here are three numbers that make a fact family: **2**, **4**, and **6**. **Two** and **four** are the **parts**. **Six** is the **whole**.*
- *Trace the addition sentence **2 + 4 = 6**.*
- *Now write the related addition sentence. Switch the order of the parts. Write **4 + 2 = 6**.*
- *Trace the subtraction sentence **6 – 2 = 4**.*
- *Now write the related subtraction sentence. Start with the whole, which is always the biggest number of the three. Look at the previous subtraction sentence. Switch the order of the parts. Write **6 – 4 = 2**.*

Repeat the process with the fact family **2, 7, 9**.

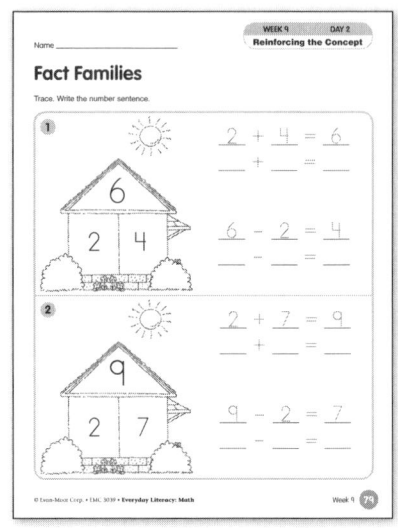

Day 2 activity

Day 3
SKILLS

Algebra
- Determine the unknown whole number in an equation or word problem
- Understand the relationship between addition and subtraction

Literacy

Oral Language Development
- Respond orally to simple questions
- Use mathematical terms

Comprehension
- Make connections using illustrations, prior knowledge, or real-life experiences

Applying the Concept

Distribute the Day 3 activity and pencils. Then guide students through the activity by saying:

- *The fact family house is missing a number.*
- *We know **8** is the whole and **3** is one of the parts. The first addition sentence says this.*
- *What can we add to **3** to equal **8**? (5) That's right, **3 + 5 = 8**. **Three** is one part and **5** is the other. Write the number **5** in the empty spot in the house. Also write **5** to finish the addition sentence **3 + 5 = 8**.*
- *Look at the next addition sentence. It's also missing one of the parts. What can we add to **5** so that it equals **8**? (3) Yes, **3** is a member of the fact family **3, 5, 8**. You know **3 + 5 = 8**; that is the addition sentence you just made. If you know **3 + 5 = 8**, then you know **5 + 3 = 8**. Write **3** to finish the addition sentence **5 + 3 = 8**.*
- *Now look at the subtraction sentence **8 – 3 = ___**. What number is missing? (5) Yes, **8 – 3 = 5**. Write the number **5** on the line. This sentence has all the members of the fact family **3, 5, 8**.*
- *Now look at the related subtraction sentence **___ – 5 = 3**. What number is missing? (8) Yes, **8 – 5 = 3**. Write the number **8** on the line. Now we have all four sentences for the fact family **3, 5, 8**.*

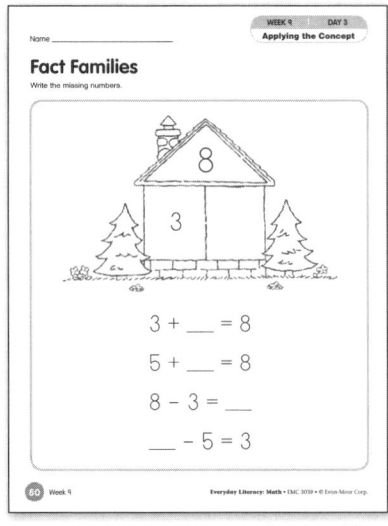

Day 3 activity

Week 9

Everyday Literacy: Math • EMC 3039 • © Evan-Moor Corp.

Day 4 SKILLS

Algebra
- Understand the relationship between addition and subtraction

Literacy

Oral Language Development
- Respond orally to simple questions
- Use mathematical terms

Comprehension
- Make connections using illustrations, prior knowledge, or real-life experiences

Extending the Concept

Distribute the Day 4 activity and pencils. Then guide students through the activity by saying:

- *The fact family house is empty, and we are going to fill it with the numbers in the cloud. The numbers 3, 6, and 9 are a **fact family**.*

- *Which numbers are the parts?* (the two smaller numbers, 3 and 6) *Write the two parts in the boxes.*

- *Which number is the whole?* (9) *Yes, 9 is the whole. Write 9 on the roof of the house.*

- *Now it is time to write the four related number sentences. Let's start with the two addition sentences. First add the two parts, 3 and 6. Below the house, write the number 3 on the first line. Then trace the plus sign. Now write the number 6. Next, trace the equal sign. What is the sum?* (9) *Write the number 9 on the last line.*

- *Write the next addition sentence. Remember: Just switch the order of the addends, 3 and 6.* (6 + 3 = 9)

- *Next, write a subtraction sentence. Start with the whole, which is 9. Trace the minus sign. Then subtract one of the parts.* (9 – 3 or 9 – 6)

- *Write the last subtraction sentence. Start with the whole again, 9. Trace the minus sign. Then subtract the other part.* (9 – 6 or 9 – 3)

Day 4 activity

Day 5 SKILLS

Algebra
- Understand the relationship between addition and subtraction

Mathematical Thinking and Reasoning
- Use math to solve problems

Home–School Connection p. 82
Spanish version available (see p. 2)

Hands-on Math Activity

Reinforce this week's math concept with the following hands-on activity:

Materials: letter-size envelopes (about 3" x 7"), 3" x 5" index cards

Activity: Give each student an index card and an envelope. Have them open the envelope so the flap is sticking up. Explain that this is a fact family house, with the flap as the roof. Assign each student a fact family, such as **2, 3, 5**; **3, 4, 7**; **2, 6, 8**; or **4, 6, 10**.

Have students label their envelope with their fact family, with the two parts on the bottom section of the envelope and the biggest number, or the whole, on the inside flap.

Next, on the index card, have students write two addition sentences and two subtraction sentences that go with their fact family. For example, the fact family **2, 6, 8** would feature the number sentences 2 + 6 = 8, 6 + 2 = 8; 8 – 2 = 6, 8 – 6 = 2. Have students tuck their number sentences inside the envelope.

On a bulletin board, tack up the fact family houses by the flaps to make a fact family neighborhood.

© Evan-Moor Corp. • EMC 3039 • **Everyday Literacy: Math** Week 9

Name _____

WEEK 9 | DAY 1
Confirming Understanding

Fact Families

$\underline{2} + \underline{3} = \underline{5}$

$\underline{} + \underline{} = \underline{}$

$\underline{5} - \underline{3} = \underline{2}$

$\underline{} - \underline{} = \underline{}$

78 Week 9

Everyday Literacy: Math • EMC 3039 • © Evan-Moor Corp.

Name _____

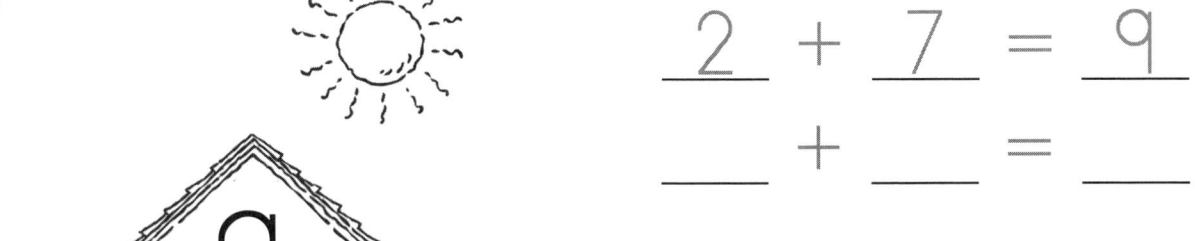

WEEK 9 | DAY 2
Reinforcing the Concept

Fact Families

Trace. Write the number sentence.

1

6
2 4

2 + 4 = 6
__ + __ = __

6 − 2 = 4
__ − __ = __

2

9
2 7

2 + 7 = 9
__ + __ = __

9 − 2 = 7
__ − __ = __

© Evan-Moor Corp. • EMC 3039 • **Everyday Literacy: Math**   Week 9 79

Name _____

WEEK 9 | DAY 3
Applying the Concept

Fact Families

Write the missing numbers.

3 + ___ = 8

5 + ___ = 8

8 − 3 = ___

___ − 5 = 3

WEEK 9 | DAY 4
Extending the Concept

Name _____

Fact Families

Write the sentences.

3, 6, 9

___ + ___ = ___

___ + ___ = ___

___ − ___ = ___

___ − ___ = ___

Name _____

What I Learned

What to Do
Have your child look at the picture below and tell you about the relationship among the three numbers. (The three numbers are related: 5 is the whole; 2 and 3 are the parts. They make addition and subtraction sentences.) Then have your child trace and complete the number sentences.

WEEK 9

Home–School Connection

Math Concept: Addition and subtraction have an inverse relationship.

To Parents
This week your child learned about fact families.

2 + 3 = 5

__ + __ = __

5 − 3 = 2

__ − __ = __

What to Do Next
Give your child three numbers that make a fact family, such as 4, 5, 9. Have him or her write the numbers on one side of an index card, and on the back, write the four related number sentences.

WEEK 10

Concept
Word problems may be solved in different ways.

What's the Problem?

Math Objective:
To help students learn strategies to solve word problems

Math Vocabulary:
count back, count on, difference, word problem

Day 1 SKILLS

Algebra
- Use more than one strategy to solve a problem

Literacy

Oral Language Development
- Respond orally to simple questions
- Use mathematical terms

Comprehension
- Make connections using illustrations, prior knowledge, or real-life experiences
- Answer questions about key details in a text read aloud

Introducing the Concept

Ask students to listen to this word problem:

Three crabs are in the sand. Eight crabs are swimming. How many crabs are there? We need to add 3 + 8.

Write 3 + 8 = ? on the board. Show a number line; point to the number **8**. Then say:

- One way to solve this addition problem is to **count on**. You start with **8** because it is bigger than the number **3**. Then you count on three numbers: **9, 10, 11**. Yes, 3 + 8 = 11.

- Now count on to find **10 + 2**. Start with **10**. Count on two numbers. (11, 12) That's right, 10 + 2 = 12.

Listening to the Story

Distribute the Day 1 activity. Say: *Look at the picture and listen as I read about a teacher who used "count on" to solve an addition problem.*

The first graders were having a party. Everyone was bringing something yummy to share. One student brought in **12** cookies. Someone else brought in **3** cookies. Mrs. Lemar counted the cookies, starting with **12** and counting on three more: **13, 14, 15**. There were **15** cookies but **18** students. Would everyone get a cookie? Mrs. Lemar hoped more goodies were on the way!

Confirming Understanding

Distribute pencils. Develop the math concept by asking questions about the story. Ask:

- *How did Mrs. Lemar figure out the answer to the addition problem?* (She counted on.)

- *Count on to figure out how many cookies there were. Start with* **12** *and count on three numbers. Write* **13, 14, 15**.

- *Finish the addition sentence. Write how many cookies there were in all.* (15)

Day 1 picture

© Evan-Moor Corp. • EMC 3039 • *Everyday Literacy: Math*

Week 10   83

Day 2 SKILLS

Algebra
- Use more than one strategy to solve a problem

Literacy

Oral Language Development
- Respond orally to simple questions
- Use mathematical terms

Comprehension
- Make connections using illustrations, prior knowledge, or real-life experiences
- Answer questions about key details in a text read aloud

Reinforcing the Concept

Reread the Day 1 story. Then reinforce this week's math concept by discussing the story. Say:

In our story, a teacher counted on from 12. How many cookies were there in all? (15)

Distribute the Day 2 activity and pencils. Say:

- Point to box 1. **There were 6 cookies in a bag. Two more cookies were on a plate. How many cookies were there in all?** *Count on. Start with 6. Count on two more. Write 7, 8 on the lines. There were 8 cookies in all.*

- Point to box 2. **There were 4 chocolates in a box. Three more chocolates were on a plate. How many chocolates were there in all?** *Start with 4 and count on three more. Write 5, 6, 7 on the lines.*

- Point to box 3. **Five sticks of gum were in a pack. Four more sticks were loose. How many sticks of gum were there in all?** *Start with 5 and count on four more. Write 6, 7, 8, 9 on the lines.*

- Point to box 4. **Ten candy sticks were together. Five sticks were loose. How many sticks were there in all?** *Start with 10 and count on five more. Write 11, 12, 13, 14, 15 on the lines.*

Day 2 activity

Day 3 SKILLS

Algebra
- Use more than one strategy to solve a problem

Literacy

Oral Language Development
- Respond orally to simple questions
- Use mathematical terms

Comprehension
- Make connections using illustrations, prior knowledge, or real-life experiences

Applying the Concept

Introduce the activity by reviewing **count on** and introducing **count back**. Say:

*You know that you can **count on** to add. You can also **count back** to subtract.*

Use a number line to review counting back from 10. Distribute the Day 3 activity and pencils.

- Point to box 1. **There were 5 cupcakes. Two cupcakes were eaten. How many cupcakes are left?** *Start with 5 and count back two. Write 4, 3 on the lines. Write the difference in the box.* (3)

- Point to box 2. **There were 9 candies. Someone took 3 candies. How many candies are left?** *Start with 9 and count back three. Write 8, 7, 6 on the lines. Write the difference in the box.* (6) **Six** *candies are left.*

- Point to box 3. **There were 6 lollipops. Two were taken. How many lollipops are left?** *Start with 6 and count back two. Write 5, 4 on the lines. 6 – 2 = 4. Write 4 in the box.* **Four** *lollipops are left.*

- Point to box 4. **There were 7 gumdrops. Three were squished. How many gumdrops are left?** *Start with 7 and count back three. Write 6, 5, 4 on the lines. 7 – 3 = 4. Write 4 in the box.* **Four** *gumdrops are left.*

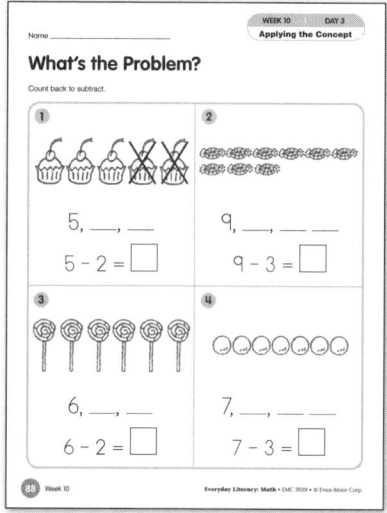

Day 3 activity

84 Week 10   Everyday Literacy: Math • EMC 3039 • © Evan-Moor Corp.

Day 4 SKILLS

Algebra
- Use more than one strategy to solve a problem

Literacy

Oral Language Development
- Respond orally to simple questions
- Use mathematical terms

Comprehension
- Make connections using illustrations, prior knowledge, or real-life experiences

Extending the Concept

Distribute the Day 4 activity and pencils. Then introduce the activity by saying:

*Listen to these problems. Decide if you will count on to **add** or count back to **subtract**.*

- Point to box 1. **Ana had 5 gumballs. Tom gave her 3 more gumballs. How many are there in all?** *Will you **count on** or **count back**?* (count on) *That's right. Start with 5. Write 6, 7, 8 on the lines. Write the addition sentence that tells this story.* (5 + 3 = 8)

- Point to box 2. **Dee had 8 brownies. She gave away 3 brownies. How many brownies are left?** *Will you **count on** or **count back**?* (count back) *Yes. Start with 8. Write 7, 6, 5 on the lines. Write the subtraction sentence that tells this story.* (8 – 3 = 5)

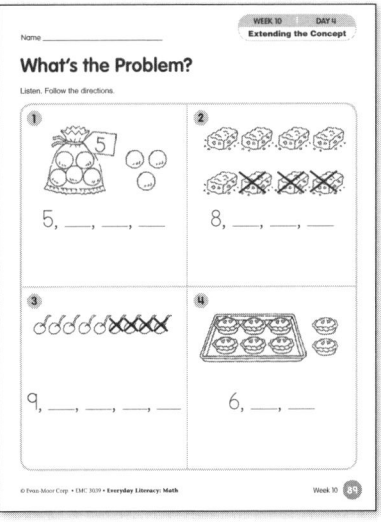

- Point to box 3. **Ted had 9 cherries. He ate 4 cherries. How many cherries are left?** *Will you **count on** or **count back**?* (count back) *Yes. Start with 9. Write 8, 7, 6, 5 on the lines. Write the subtraction sentence that tells this story.* (9 – 4 = 5)

- Point to box 4. **Pam had 6 pies. Lori gave her 2 more. How many pies are there altogether?** *Will you **count on** or **count back**?* (count on) *Yes. Start with 6. Write 7, 8 on the lines. Write the addition sentence that tells this story.* (6 + 2 = 8)

Day 4 activity

Day 5 SKILLS

Algebra
- Use more than one strategy to solve a problem

Mathematical Thinking and Reasoning
- Explore mathematical ideas through song or play

Home–School Connection p. 90
Spanish version available (see p. 2)

Circle Time Math Activity

Reinforce this week's math concept with the following circle time activity:

Materials: number line

Activity: Gather students in a circle. Teach them the chant below, pointing to the corresponding numbers on the number line.

Have students repeat the chant with you, using their right hand to jump 5 spaces to the right when counting on. When counting back, have them switch to their left hand and jump 5 spaces to the left.

*Putting together can be fun.
One good way is to count on:
5, 6, 7, 8, 9.
Move ahead on a number line!*

*But if you want to subtract,
One good way is to count back:
5, 4, 3, 2, 1.
Counting back. That's how it's done!*

© Evan-Moor Corp. • EMC 3039 • **Everyday Literacy: Math** Week 10 85

Name _____

WEEK 10 | DAY 1
Confirming Understanding

What's the Problem?

12, ____, ____, ____

12 + 3 = ☐

Week 10

Everyday Literacy: Math • EMC 3039 • © Evan-Moor Corp.

Name _____

WEEK 10 | DAY 2
Reinforcing the Concept

What's the Problem?

Count on to add.

1

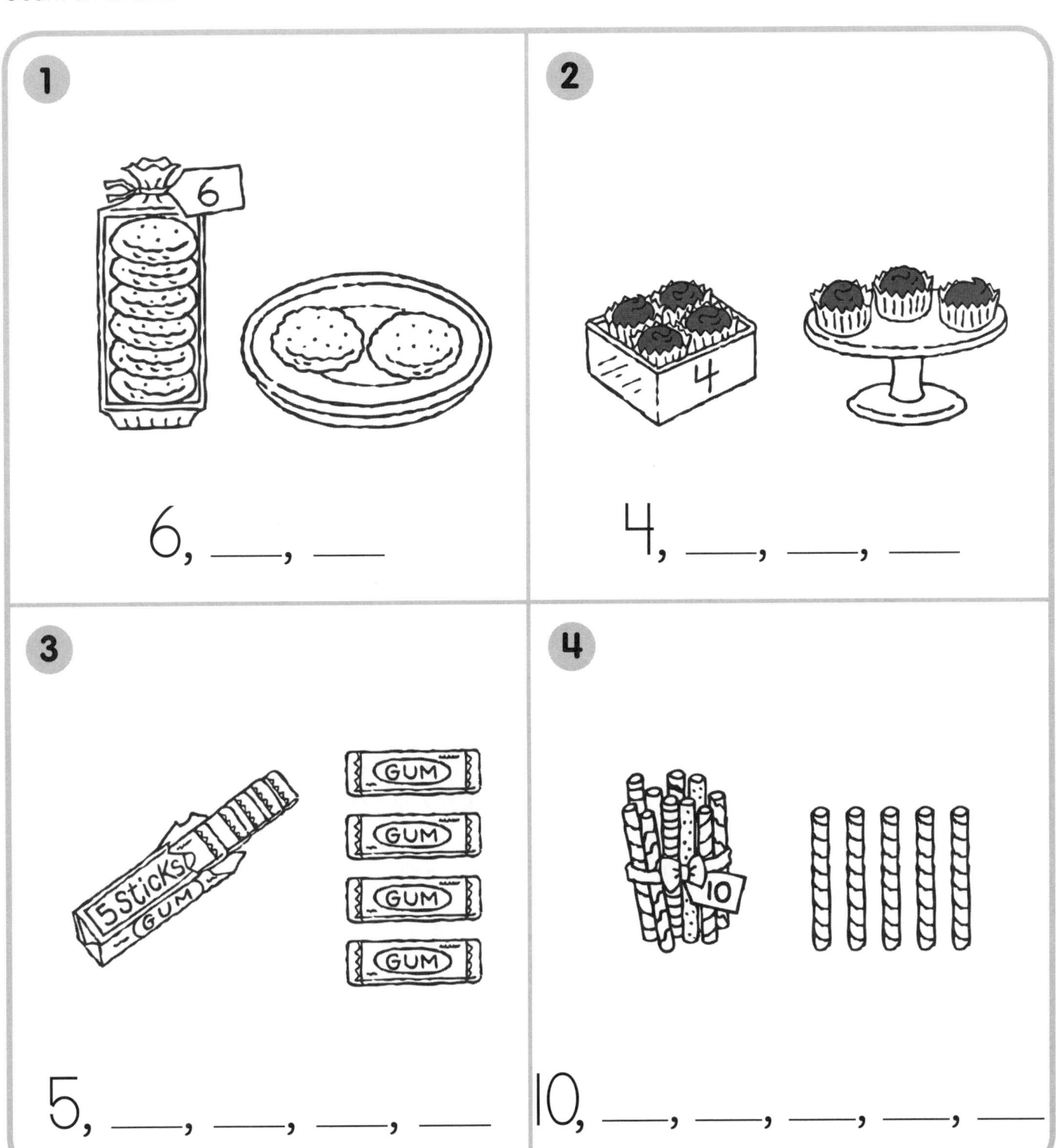

6, ___, ___

2

4, ___, ___, ___

3

5, ___, ___, ___, ___

4

10, ___, ___, ___, ___, ___

Name _____

WEEK 10 | DAY 3
Applying the Concept

What's the Problem?

Count back to subtract.

1

5, ___, ___

5 − 2 = ☐

2

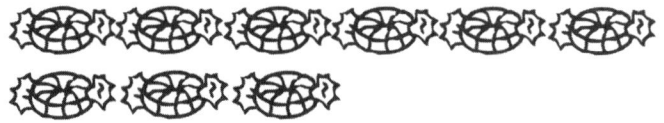

9, ___, ___ ___

9 − 3 = ☐

3

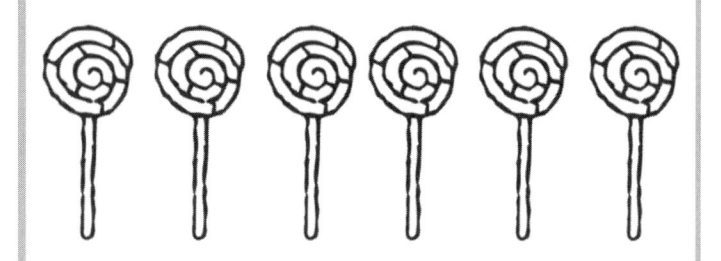

6, ___, ___

6 − 2 = ☐

4

7, ___, ___ ___

7 − 3 = ☐

88 Week 10

Everyday Literacy: Math • EMC 3039 • © Evan-Moor Corp.

Name _____

WEEK 10 | DAY 4
Extending the Concept

What's the Problem?

Listen. Follow the directions.

1

5, ___, ___, ___

2

8, ___, ___, ___

3

9, ___, ___, ___, ___

4

6, ___, ___

© Evan-Moor Corp. • EMC 3039 • **Everyday Literacy: Math**

Week 10   89

Name _____

What I Learned

What to Do
Look at the picture with your child. Then have your child use the strategy of *counting on* to add 3 cookies to the 12 cookies in the box. Have him or her start at *12* and then count on three more numbers: *13, 14, 15*. Then have your child write those numbers on the lines and complete the addition sentence.

WEEK 10

Home–School Connection

Math Concept: Word problems may be solved in different ways.

To Parents
This week your child learned to count on and count back.

12, ____, ____, ____

12 + 3 = ☐

What to Do Next
Help your child count sets of objects by using the strategy of *count on*. You might present 8 crayons in a box plus 4 loose ones, and ask your child to count on: *9, 10, 11, 12. There are 12 crayons in all!*

Week 10   Everyday Literacy: Math • EMC 3039 • © Evan-Moor Corp.

WEEK 11

Concept
Symbols can be used to show the relationship between two numbers.

Greater Than, Less Than

Math Objective:
To help students compare numbers using symbols for *greater than*, *less than*, and *equal to*

Math Vocabulary:
compare, equal to, fact, greater than, less than, symbol

Day 1 SKILLS

Algebra
- Compare numbers as *greater than*, *less than*, or *equal to*
- Use symbols that indicate a comparison

Literacy

Oral Language Development
- Respond orally to simple questions
- Use mathematical terms

Comprehension
- Make connections using illustrations, prior knowledge, or real-life experiences
- Answer questions about key details in a text read aloud

Introducing the Concept

Make cards with >, <, and =. Make two towers, one with 10 blocks and one with 7 blocks. Point out the difference in the number of blocks. Display the corresponding symbol card when introducing the symbol name. Say:

- *Which tower is bigger?* (the tower with 10 blocks) *Yes, it has 10 blocks. We say that 10 is* **greater than** *7.*
- *We can use* **symbols** *to compare the numbers 10 and 7. The symbol > is like an alligator's mouth open wide. It wants to eat the* **greater** *number. I can write* **10 > 7**. *This fact says* **10 is greater than 7**.
- *We can also say that 7 is* **less than** *10. I can write* **7 < 10**. *This fact says* **7 is less than 10**. *Now you read the fact.* (students respond)
- *If I add 3 more blocks to the 7 tower, both towers will be* **equal**. *They will both have 10 blocks. I can write* **10 = 10**. *This fact says* **10 is equal to 10**. *Now you read the fact.* (students respond)

Listening to the Story

Distribute the Day 1 activity. Say: *Look at the picture and listen as I read a story about a dog who compares two groups of treats.*

Bernie is a very big dog with a very big appetite. One day, Bernie pushed open the pantry door with his big snout. Then he used his big paw to knock over the bag of treats. The bag fell and the treats spilled into two groups. Bernie heard footsteps, so he had to act fast. Bernie was not only very big. He was also very smart. He compared the two groups of treats and quickly ate the bigger one.

Confirming Understanding

Distribute pencils. Develop the math concept by asking questions about the story. Say:

- *Look at the bottom of the page. On each line, write the number of treats in each group.* (23; 25)
- *Compare the two numbers. Which number is greater?* (25)
- *Write the symbol in the box between the two numbers. Make sure the alligator's mouth is open toward the bigger number.* (23 < 25) *Let's read the fact:* **23 is less than 25**.

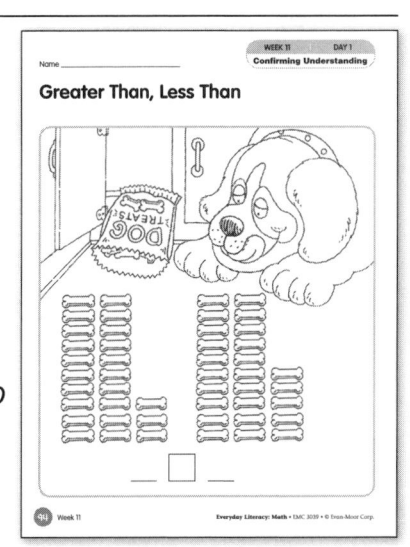

Day 1 picture

© Evan-Moor Corp. • EMC 3039 • *Everyday Literacy: Math* Week 11 **91**

Day 2 SKILLS

Algebra
- Compare numbers as *greater than*, *less than*, or *equal to*
- Use symbols that indicate a comparison

Literacy

Oral Language Development
- Respond orally to simple questions
- Use mathematical terms

Comprehension
- Make connections using illustrations, prior knowledge, or real-life experiences
- Answer questions about key details in a text read aloud

Reinforcing the Concept

Reread the Day 1 story. Then reinforce this week's math concept by discussing the story. Say:

*Our story was about two groups of doggie treats. One had **25** treats and one had **23** treats. Which number is **greater**?* (25)

Distribute the Day 2 activity and crayons. Say:

- *Now we are going to compare pairs of numbers to see which one is **greater**. Point to box 1. Which number is greater, **5** or **4**?* (5) *That's right. Color the paw with the **5**.*

- *Point to box 2. Which number is greater, **8** or **6**?* (8) *That's right. Color the paw with the **8**.*

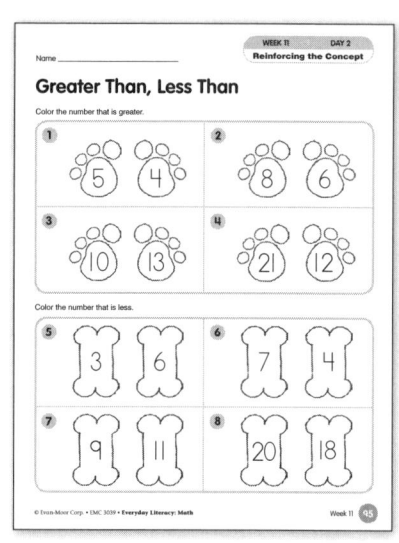

Repeat the process for boxes 3 and 4. Then say:

- *Now let's compare pairs of numbers to see which one is **less**. Point to box 5. Which number is less, **3** or **6**?* (3) *That's right. Color the bone with the **3**.*

- *Point to box 6. Which number is less, **7** or **4**?* (4) *Yes. Color the bone with the **4**.*

Repeat the process for boxes 7 and 8.

Day 2 activity

Day 3 SKILLS

Algebra
- Compare numbers as *greater than*, *less than*, or *equal to*
- Use symbols that indicate a comparison

Literacy

Oral Language Development
- Respond orally to simple questions
- Use mathematical terms

Comprehension
- Make connections using illustrations, prior knowledge, or real-life experiences

Applying the Concept

Distribute the Day 3 activity and pencils. Then guide students through the activity by saying:

- *We are going to compare pairs of numbers. Point to number 1. What are the two numbers?* (41 and 41) *Is 41 **greater than**, **less than**, or **equal to** 41?* (equal to) *Circle the symbol for **equal to**.*

- *Point to number 2. What are the two numbers?* (24 and 26) *Is 24 **greater than**, **less than**, or **equal to** 26?* (less than) *Circle the symbol for **less than**.*

- *Point to number 3. What are the two numbers?* (42 and 23) *Is 42 **greater than**, **less than**, or **equal to** 23?* (greater than) *Circle the symbol for **greater than**.*

- *Point to number 4. What are the two numbers?* (30 and 40) *Is 30 **greater than**, **less than**, or **equal to** 40?* (less than) *Circle the symbol for **less than**.*

Repeat the process for numbers 5 and 6.

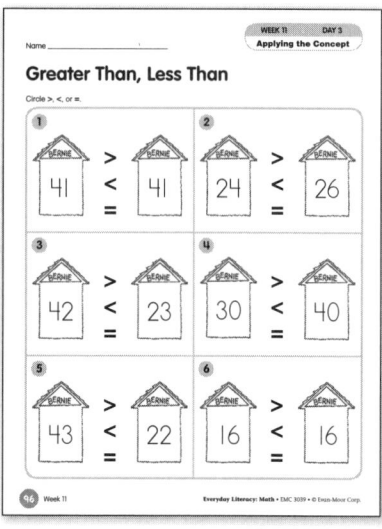

Day 3 activity

---

92   Week 11   Everyday Literacy: Math • EMC 3039 • © Evan-Moor Corp.

Day 4
SKILLS

Algebra
- Compare numbers as *greater than*, *less than*, or *equal to*
- Use symbols that indicate a comparison

Literacy

Oral Language Development
- Respond orally to simple questions
- Use mathematical terms

Comprehension
- Make connections using illustrations, prior knowledge, or real-life experiences

Extending the Concept

Distribute the Day 4 activity and pencils. Then introduce the activity by saying:

*You can **count on** to add. Remember to start with the known number and count forward.*

- *Point to row 1. **The first bag has 5 doggie treats. There are 3 more loose treats. How many doggie treats are there in all?** Start with **5** and count on three more: **6, 7, 8**. There are **8** treats in all. Write the number **8** in the box.*

- ***The second bag has 5 doggie treats. There are 4 more loose ones. How many doggie treats are there in all?** Start with **5** and count on four more: **6, 7, 8, 9**. There are **9** treats in all. Write the number **9** in the box.*

- *Now compare the two numbers: **8** and **9**. Is 8 **greater than**, **less than**, or **equal to** 9?* (less than) *Circle the symbol for **less than**.*

Repeat the process for rows 2 and 3.

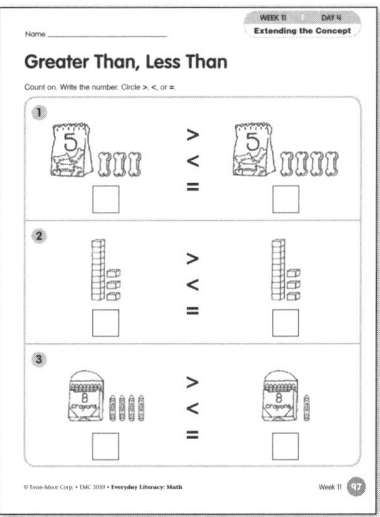

Day 4 activity

Day 5
SKILLS

Algebra
- Compare numbers as *greater than*, *less than*, or *equal to*
- Use symbols that indicate a comparison

Mathematical Thinking and Reasoning
- Select and use various types of reasoning and methods of proof

Home–School Connection p. 98
Spanish version available (see p. 2)

Hands-on Math Activity

Reinforce this week's math concept with the following hands-on activity:

Materials: green pipe cleaners, small pompoms

Activity: Divide students into pairs: Student A and Student B. Give each pair a green pipe cleaner and about 20 pompoms.

Have the students bend their pipe cleaners into a V, to form an alligator mouth. Have them turn the V on its side. Tell students that this can be either a greater than or a less than symbol, depending on how they turn it. It is in the shape of an alligator mouth to remind them that the alligator always opens its mouth toward the greater number.

Have Student A divide the 20 pompoms into two groups. Have Student B compare the two groups of pompoms, estimating the amount in each. Remind students that the alligator wants to eat the group with the greater number, so the open end of the symbol should point to the greater number.

Have Student A verify, by counting, whether Student B positioned the symbol correctly. Then ask Student A to state the number fact represented by the pompoms and symbol, for example, *8 is less than 12*.

Then have partners switch roles.

© Evan-Moor Corp. • EMC 3039 • **Everyday Literacy: Math**   Week 11

Name _____

Greater Than, Less Than

WEEK 11 | DAY 1
Confirming Understanding

___ ___ ___

Week 11 — Everyday Literacy: Math • EMC 3039 • © Evan-Moor Corp.

Name _____

WEEK 11 | DAY 2
Reinforcing the Concept

Greater Than, Less Than

Color the number that is greater.

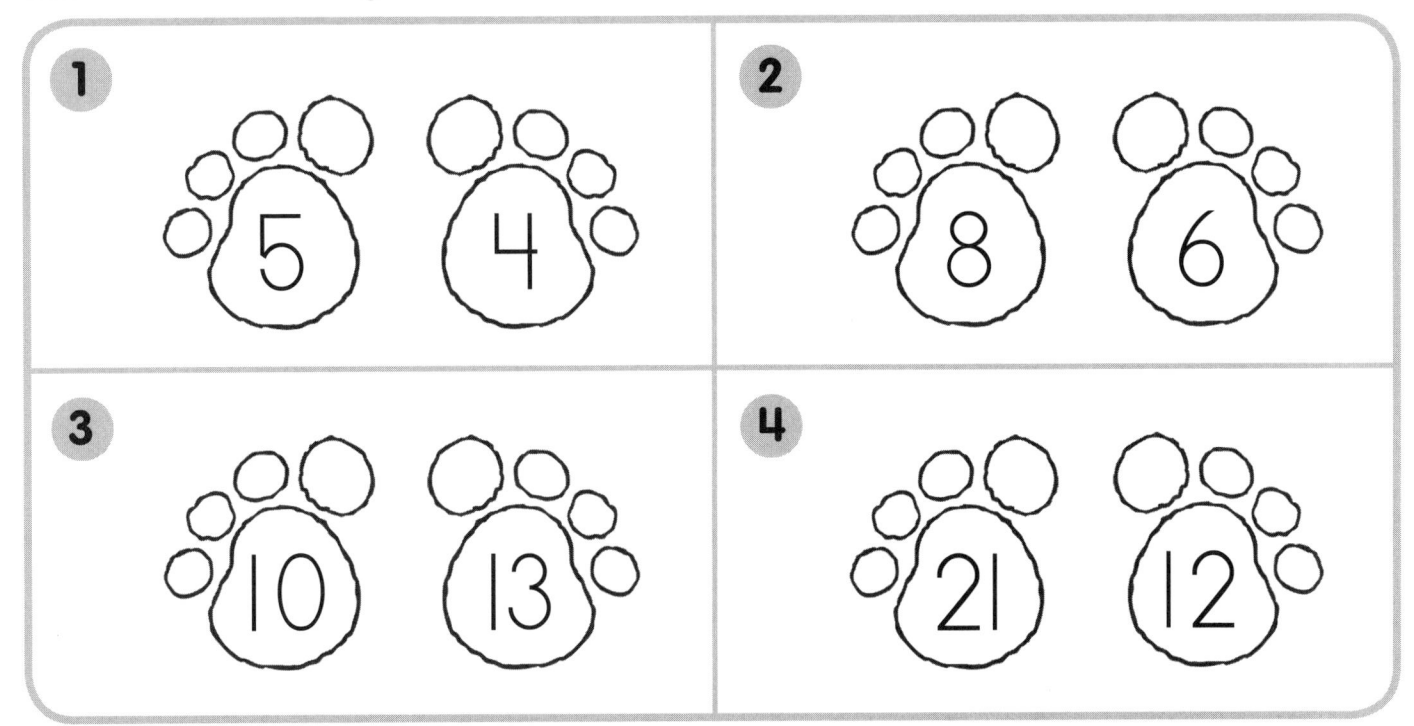

1. 5   4
2. 8   6
3. 10   13
4. 21   12

Color the number that is less.

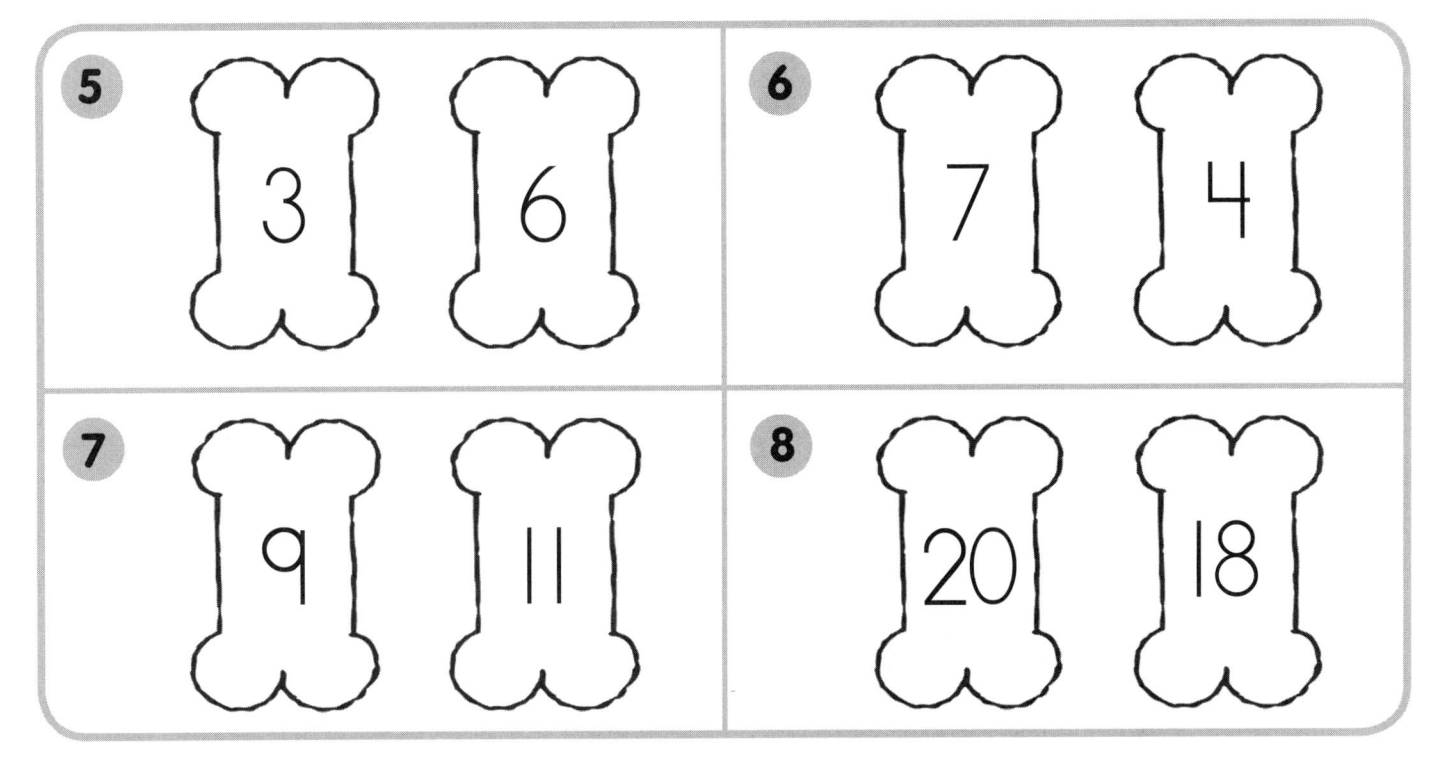

5. 3   6
6. 7   4
7. 9   11
8. 20   18

© Evan-Moor Corp. • EMC 3039 • *Everyday Literacy: Math*   Week 11

Name _____

WEEK 11 | DAY 3
Applying the Concept

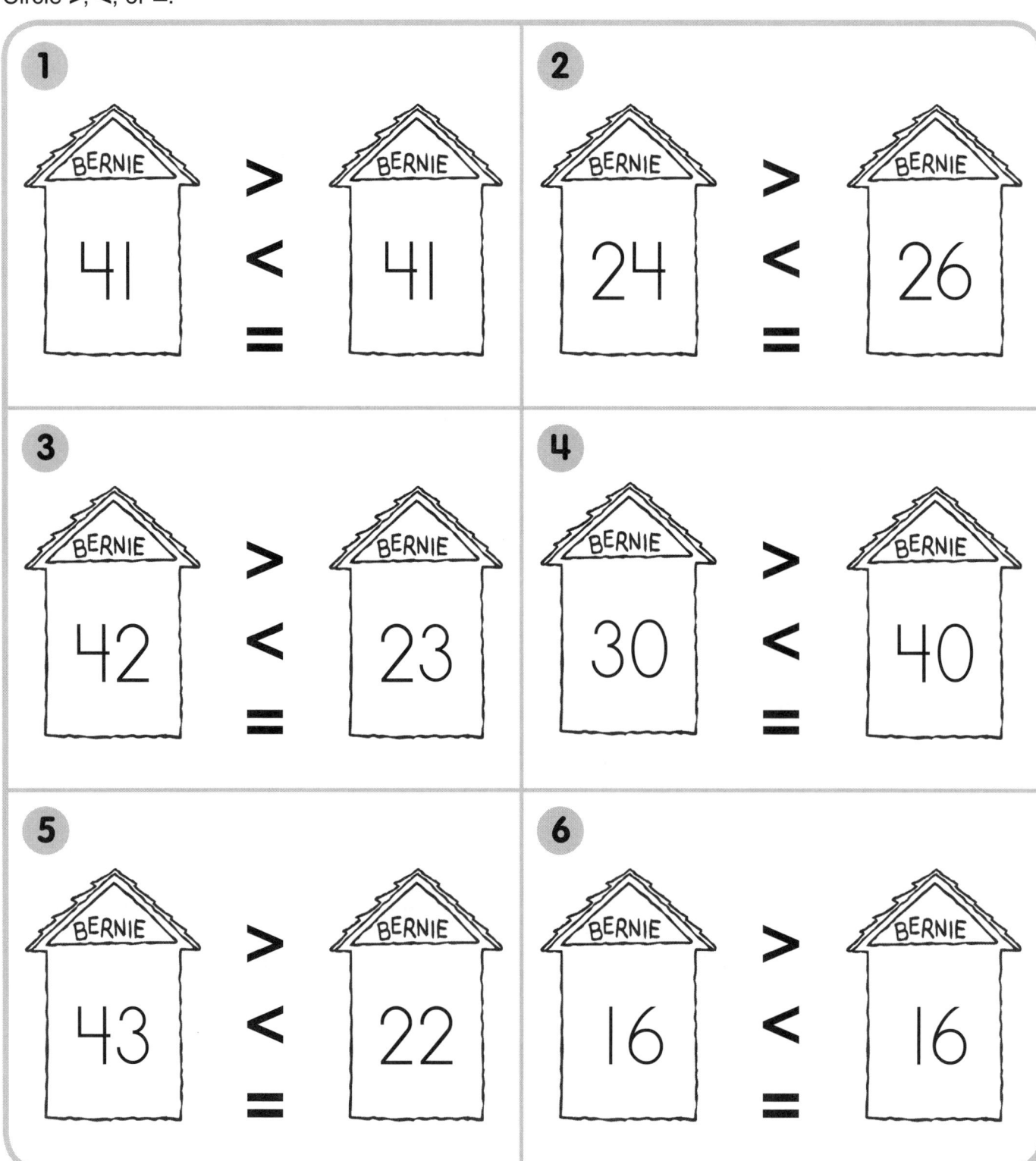

Greater Than, Less Than

Circle >, <, or =.

1

41 > < = 41

2

24 > < = 26

3

42 > < = 23

4

30 > < = 40

5

43 > < = 22

6

16 > < = 16

Greater Than, Less Than

Count on. Write the number. Circle >, <, or =.

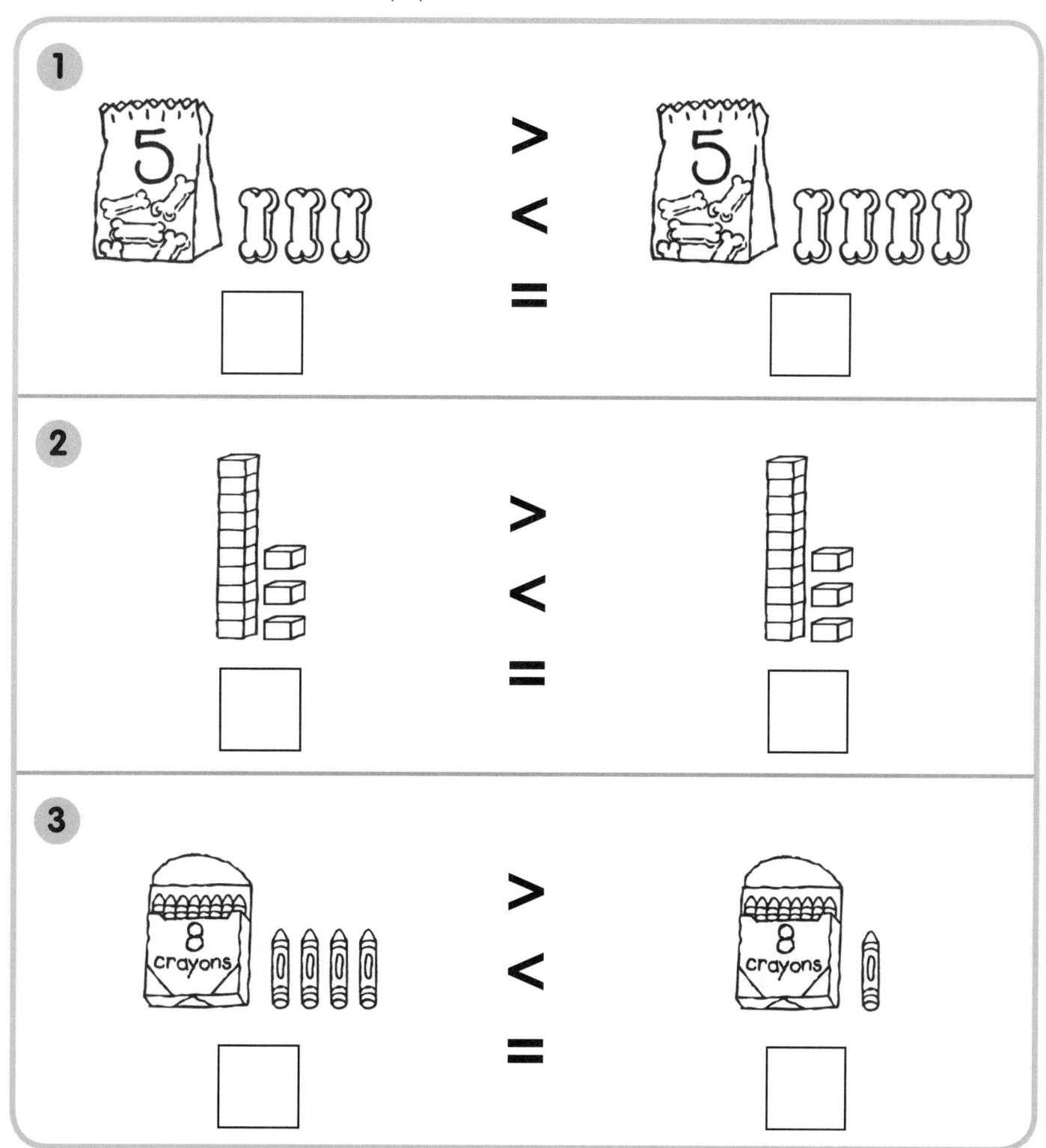

1.

\> < =

2.

\> < =

3.

\> < =

Name _____

What I Learned

WEEK 11

Home–School Connection

What to Do
Look at the picture below with your child. Then have your child compare the two groups of dog treats by using the words *greater than* or *less than*. Next, have your child write *23 < 25* on the lines at the bottom of the page and tell you what it means.

Math Concept: Symbols can be used to show the relationship between two numbers.

To Parents
This week your child learned to compare numbers using symbols.

____ ▢ ____

What to Do Next
On index cards, write the symbols >, < , and =. Then have your child make groups of mini pretzel sticks (or other snacks) that vary in number and compare the two groups using >, <, and =.

Week 11   Everyday Literacy: Math • EMC 3039 • © Evan-Moor Corp.

WEEK 12

Concept
Objects may be described using the names of shapes.

Basic Shapes

Math Objective:
To help students identify two-dimensional shapes and their attributes

Math Vocabulary:
circle, corner, oval, rectangle, side, square, triangle

Day 1 SKILLS

Geometry
- Recognize simple shapes regardless of size or orientation

Literacy

Oral Language Development
- Respond orally to simple questions
- Use the names of shapes

Comprehension
- Make connections using illustrations, prior knowledge, or real-life experiences
- Answer questions about key details in a text read aloud

Introducing the Concept

Display the following shapes: two circles of different sizes, two squares of different colors, a triangle, an oval, and a rectangle. Tell students you will ask them questions about the shapes and they will show a thumbs up to answer **yes** and a thumbs down to answer **no**. Point to the shape as you ask a question, such as:

- *Is a circle round?* (yes) *Does a triangle have four sides?* (no)
- *When I flip the rectangle, is it still a rectangle?* (yes)
- *If a square is yellow, is it still a square?* (yes)
- *Does a rectangle have corners?* (yes)
- *Does a square have two long sides and two short sides?* (no)
- *Does an oval have corners?* (no)

Listening to the Story

Distribute the Day 1 activity. Say: *Listen as I read a story about a boy who sees shapes on a bus. Point to the shapes as I read about them.*

At first, Roman thought he was looking at a regular bus. Then he noticed the shapes. There were **big circles** and **small circles**. There were **squares** and **rectangles**, including one **long rectangle** along the side of the bus. Even the lines on the street were **rectangles**. It seemed that the more he looked, the more shapes Roman saw!

Confirming Understanding

Distribute crayons. Develop the math concept by asking questions about the story. Ask:

- *What are the shapes without sides and corners?* (circles and ovals) *Find the two small* **circles** *and color them brown.*
- *How many shapes have four sides that are the same?* (1) *What is this shape called?* (square) *Find a* **square** *and color it red.*
- *How many* **rectangles** *are on the bus?* (7) *What is the same about a rectangle and a square?* (Both have 4 corners and 4 sides.) *Color one rectangle green.*

Day 1 picture

© Evan-Moor Corp. • EMC 3039 • Everyday Literacy: Math

Week 12 99

Day 2 SKILLS

Geometry
- Recognize simple shapes regardless of size or orientation

Literacy

Oral Language Development
- Respond orally to simple questions
- Use the names of shapes

Comprehension
- Make connections using illustrations, prior knowledge, or real-life experiences
- Answer questions about key details in a text read aloud

Reinforcing the Concept

Reread the Day 1 story. Then reinforce this week's math concept by discussing the story. Say:

Our story was about a bus with shapes. What shape did it have along the side? (rectangle)

Distribute the Day 2 activity and crayons. Say:

- *Point to number 1. Let's read this word together: **triangle**. How many sides does a triangle have?* (3) *Find the shape that has **3 sides**. Draw a line from the word **triangle** to the triangle shape.*

- *Point to number 2. Let's read this word together: **square**. How many sides does a square have?* (4) *Find the shape that has **4 sides**. Draw a line from the word **square** to the square shape. Make sure the square has **4 equal** sides.*

- *Point to number 3. Let's read this word together: **circle**. How many sides does a circle have?* (none) *Find the shape that does not have sides. Draw a line from the word **circle** to the circle shape.*

- *Point to number 4. Let's read this word together: **rectangle**. What does a rectangle look like?* (2 short sides, 2 long sides) *Find the shape that looks like this. Draw a line from the word **rectangle** to the rectangle shape.*

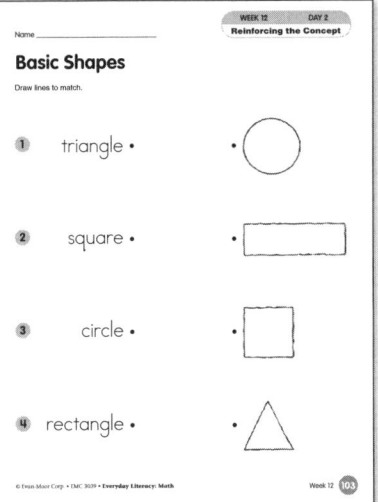

Day 2 activity

Day 3 SKILLS

Geometry
- Recognize simple shapes regardless of size or orientation

Literacy

Oral Language Development
- Respond orally to simple questions
- Use the names of shapes

Comprehension
- Make connections using illustrations, prior knowledge, or real-life experiences

Applying the Concept

Distribute the Day 3 activity and crayons. Then guide students through the activity by saying:

- *Point to box 1. Look at these shapes. Color the shapes that do not have corners.* (students color) *Which shapes did you color?* (circle and oval)

- *Point to box 2. Look at these shapes. Color the shape that has **4 equal sides**.* (students color) *Which shape did you color?* (square)

- *Point to box 3. Look at these shapes. Color the shapes that have **3 corners**.* (students color) *Which shapes did you color?* (triangles)

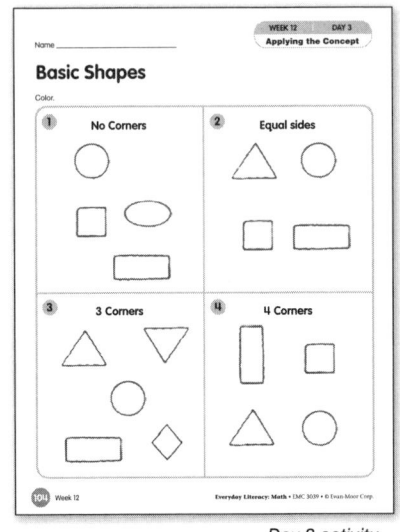

- *Point to box 4. Look at these shapes. Color the shapes that have **4 corners**.* (students color) *Which shapes did you color?* (square and rectangle)

Day 3 activity

---

100 Week 12

Everyday Literacy: Math • EMC 3039 • © Evan-Moor Corp.

Day 4 SKILLS

Geometry
- Recognize simple shapes regardless of size or orientation

Literacy

Oral Language Development
- Respond orally to simple questions
- Use the names of shapes

Comprehension
- Make connections using illustrations, prior knowledge, or real-life experiences

Extending the Concept

Distribute the Day 4 activity and crayons. Then introduce the activity by saying:

These pictures show objects made of shapes. Let's count how many of each shape there are.

Point to number 1. It shows a boat made of shapes. What shapes do you see? (rectangles and triangles)

- *How many rectangles are there?* (2) *Write the number **2** by the word **rectangles**.*
- *How many triangles are there?* (4) *Write the number **4** by the word **triangles**.*

Point to number 2. It shows a house made of shapes. What shapes do you see? (triangles, squares, rectangles, circles)

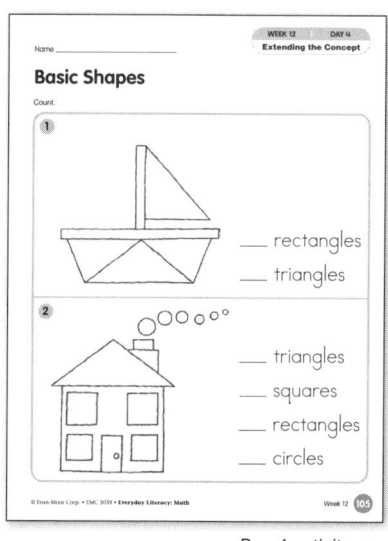

- *How many triangles are there?* (2) *Write the number **2** by the word **triangles**.*
- *How many squares are there?* (4) *Write the number **4** by the word **squares**.*
- *How many rectangles are there?* (3) *Write the number **3** by the word **rectangles**.*
- *How many circles are there?* (7) *Write the number **7** by the word **circles**.*

Day 4 activity

Day 5 SKILLS

Geometry
- Recognize simple shapes regardless of size or orientation

Mathematical Thinking and Reasoning
- Explore mathematical ideas through song or play

Home–School Connection p. 106
Spanish version available (see p. 2)

Circle Time Math Activity

Reinforce this week's math concept with the following circle time activity:

Materials: circles, rectangles, squares, triangles, and ovals cut from construction paper (make enough so each student gets one shape); index cards with shape names on them

Activity: Gather in a large outdoor space. Have students stand in a circle. Distribute one shape to each student. Explain that students will play a version of "Simon Says," using shapes.

Choose an index card with a shape name. Think of a silly movement for students to perform. Have students who are holding that particular shape perform the movement a certain number of times. For example, say, *Simon says all rectangles hop backward seven times.* Then have the rectangles return to their place in the circle.

Other movements may include jumping jacks, hopping on one foot, twisting, and touching toes.

© Evan-Moor Corp. • EMC 3039 • *Everyday Literacy: Math*   Week 12   101

Name _____

WEEK 12 | DAY 1
Confirming Understanding

Basic Shapes

Basic Shapes

Draw lines to match.

1. triangle •   • ◯

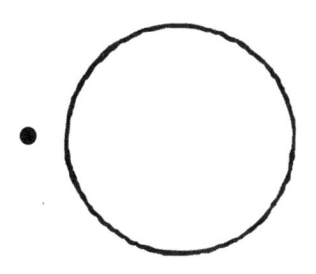

2. square •   • ▭

3. circle •   • ☐

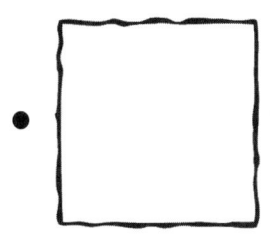

4. rectangle •   • △

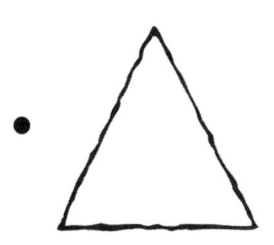

Name _____

WEEK 12 | DAY 3
Applying the Concept

Basic Shapes

Color.

| 1 No Corners | 2 Equal sides |
| 3 3 Corners | 4 4 Corners |

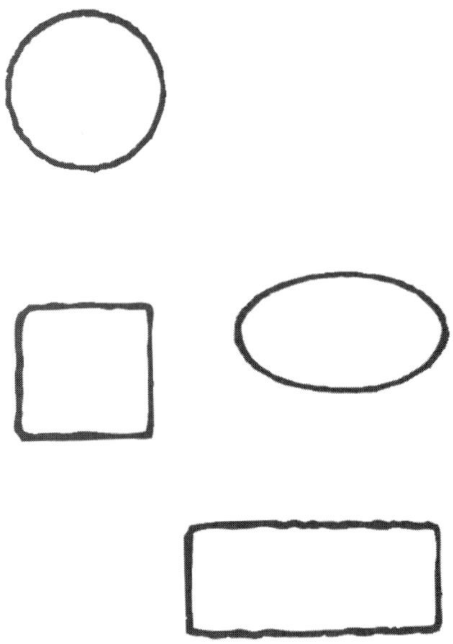

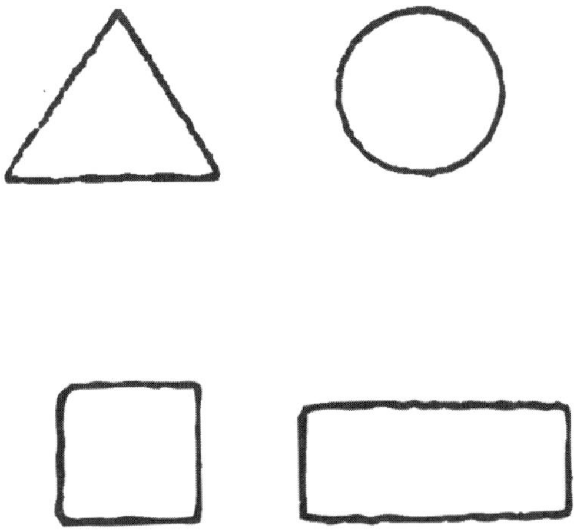

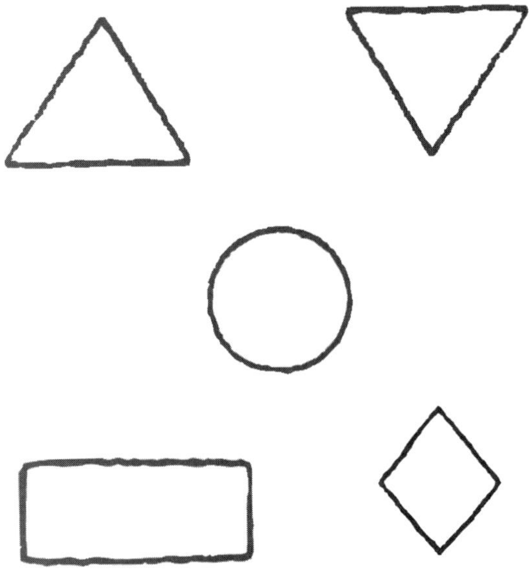

Name _____

WEEK 12 | DAY 4
Extending the Concept

Basic Shapes

Count.

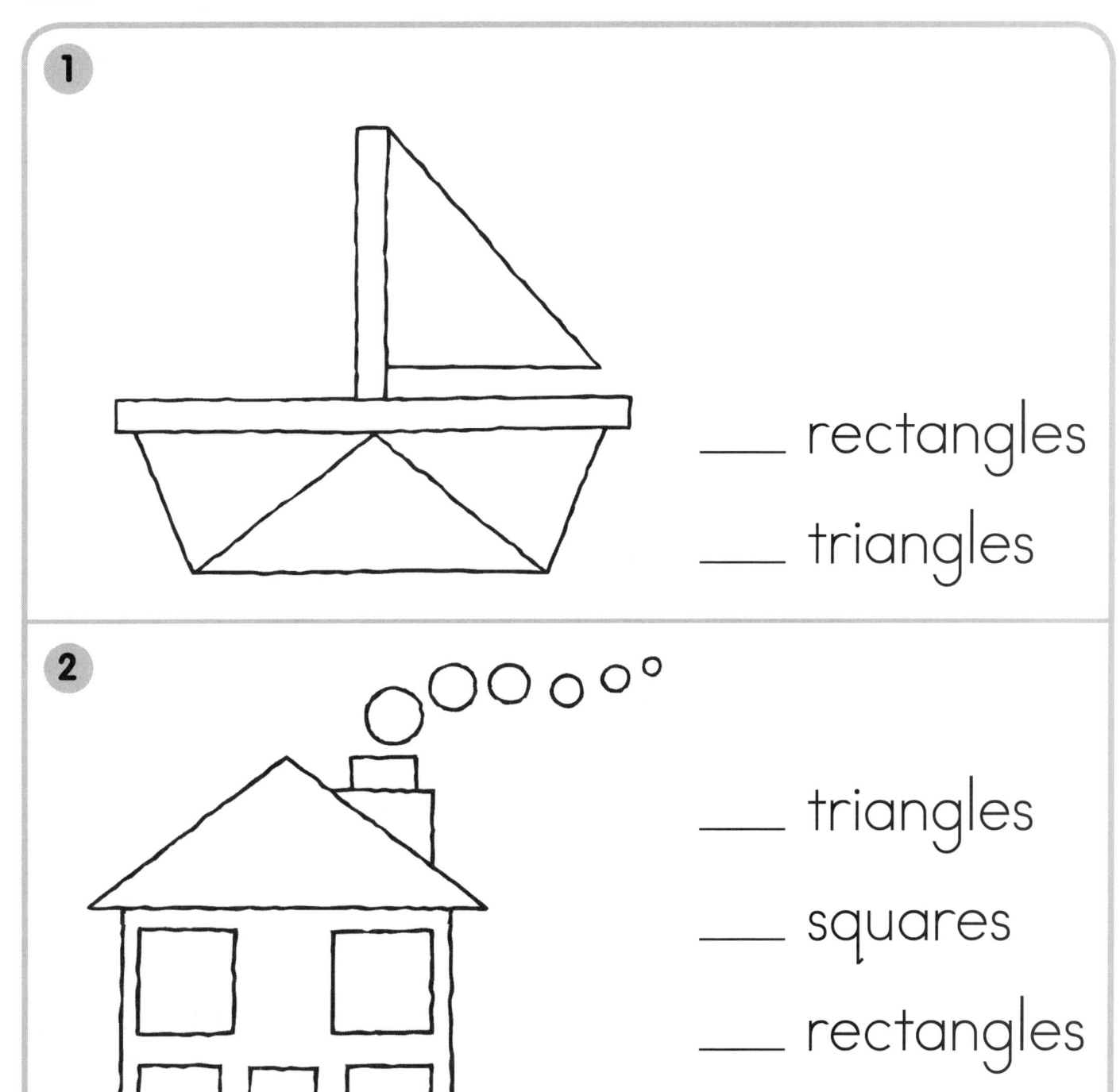

1

___ rectangles

___ triangles

2

___ triangles

___ squares

___ rectangles

___ circles

Name _____

What I Learned

What to Do
Have your child look at the picture below and point out the shapes on the bus. Then have your child count the shapes and write the number of each on the lines below.

WEEK 12

Home–School Connection

Math Concept: Objects may be described using the names of shapes.

To Parents
This week your child learned to identify basic shapes.

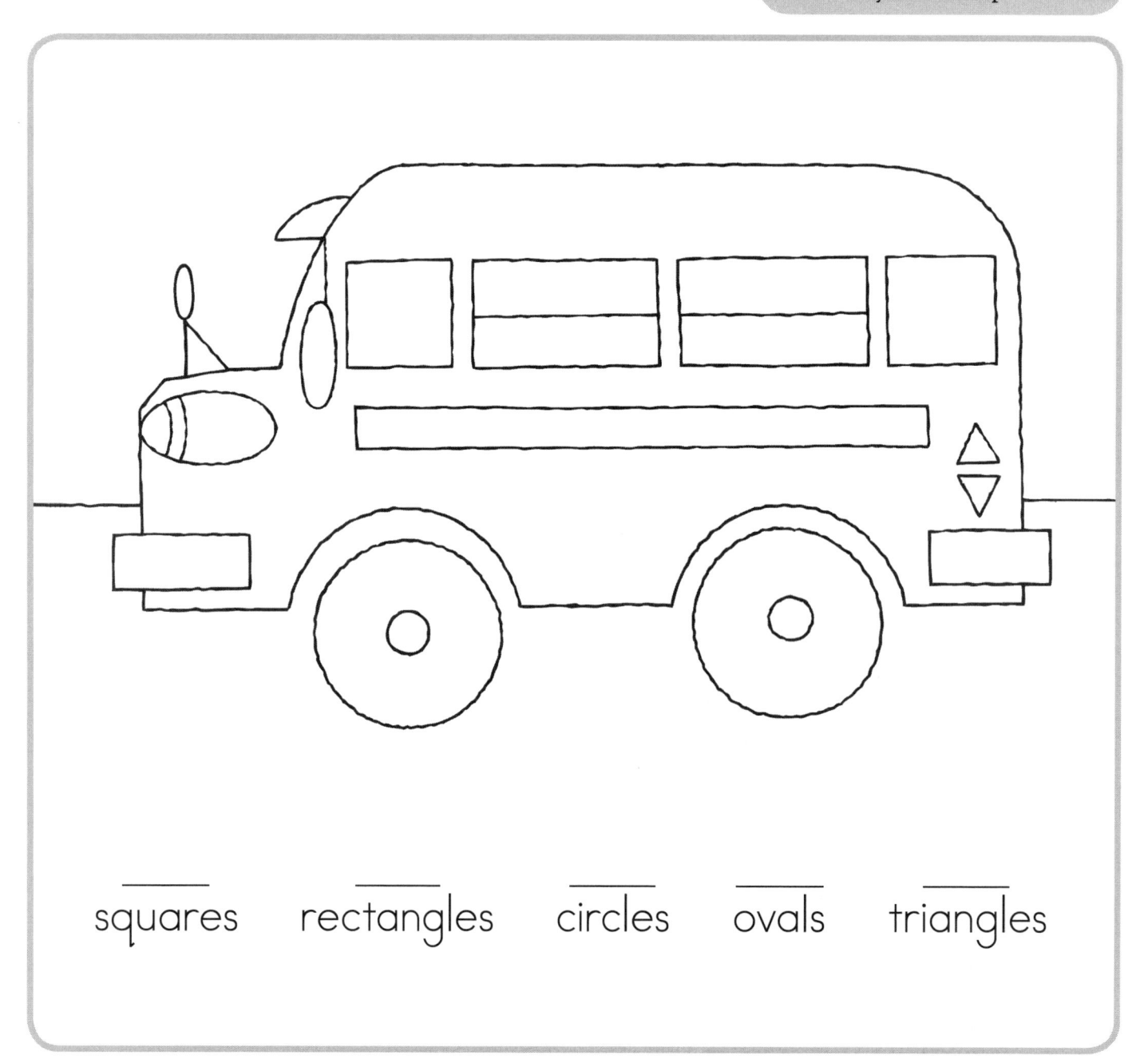

___ squares   ___ rectangles   ___ circles   ___ ovals   ___ triangles

What to Do Next
Help your child look through magazines to find and highlight items in the shapes of circles, rectangles, squares, triangles, and ovals.

106   Week 12   Everyday Literacy: Math • EMC 3039 • © Evan-Moor Corp.

WEEK 13

Concept

Some shapes can be divided into halves and fourths.

Fractions

Math Objective:
To help students understand the concept of halves and fourths

Math Vocabulary:
divide, equal parts, fourths, fractions, half, halves, same size, whole

---

Day 1 SKILLS

Geometry
- Understand that shapes can be divided into equal parts called halves or fourths
- Understand that decomposing a shape results in smaller parts

Literacy

Oral Language Development
- Respond orally to simple questions
- Use mathematical terms

Comprehension
- Make connections using illustrations, prior knowledge, or real-life experiences
- Answer questions about key details in a text read aloud
- Make inferences and draw conclusions

Introducing the Concept

Have on hand large square, circle, and heart shapes made of colored paper. Have students watch as you fold a square in half. Then unfold it and say:

- *I **divided** one whole square into two **parts**. Look at each part. Is it smaller or bigger than the **whole** square?* (smaller)
- ***Equal parts** are the **same size**. Are these two parts equal?* (yes) *Yes, the two parts are the same size, so they are equal. Two equal parts are called **halves**. What are two equal parts called?* (halves) *Watch me write $\frac{1}{2}$ on each half.*

Repeat the process with the other shapes. Fold each shape into unequal parts, too, for students to recognize the difference. End with students folding shapes in half. Have students call the equal parts **halves**.

Listening to the Story

Distribute the Day 1 activity. Say: *Look at the picture and listen as I read a story about hungry hippos and two pies cut into parts.*

The two hungry hippos got out of the pond and waddled home. Mama Hippo had baked their favorite dinner: green grass pies. She cut each pie into two parts. Mama asked her hippos which pie they wanted to eat. The hippos looked at the pies. "We want the pie that's cut in **half**," said the hippos. "Then each of us will get the **same** amount to eat." Mama Hippo smiled. She was proud of her children for sharing!

Confirming Understanding

Distribute crayons. Develop the math concept by asking questions about the story. Say:

- *Circle the pie that the hippos want to eat. Why do they choose that pie?* (It is cut into equal parts.) *How many **parts** does the pie have?* (2)
- *Draw an **X** through the pie that is not cut into halves, or equal parts.*
- *Color the pie that has **equal parts**. Make each half a different color.*

Day 1 picture

© Evan-Moor Corp. • EMC 3039 • **Everyday Literacy: Math**

Week 13   107

Day 2 SKILLS

Geometry
- Understand that shapes can be divided into equal parts called halves or fourths
- Understand that decomposing a shape results in smaller parts

Literacy

Oral Language Development
- Respond orally to simple questions
- Use mathematical terms

Comprehension
- Make connections using illustrations, prior knowledge, or real-life experiences
- Answer questions about key details in a text read aloud

Reinforcing the Concept

Reread the Day 1 story. Then reinforce this week's math concept by discussing the story. Say:

Our story was about two hippos who shared a pie. What were the two equal parts of the pie called? (halves)

Distribute the Day 2 activity and crayons. Say:

- *Point to number 1. Is this shape divided into halves? Color the circle next to yes or no.* (no) *A shape divided into halves will have how many equal parts?* (2)

- *Point to number 2. Is this shape divided into halves? Color the circle next to yes or no.* (yes)

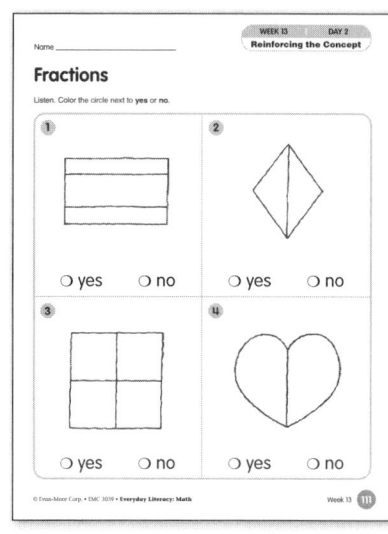

- *Point to number 3. Is this shape divided into halves? Color the circle next to yes or no.* (no)

- *Point to number 4. Is this shape divided into halves? Color the circle next to yes or no.* (yes)

Day 2 activity

Day 3 SKILLS

Geometry
- Understand that shapes can be divided into equal parts called halves or fourths
- Understand that decomposing a shape results in smaller parts

Literacy

Oral Language Development
- Respond orally to simple questions
- Use mathematical terms

Comprehension
- Make connections using illustrations, prior knowledge, or real-life experiences

Applying the Concept

Distribute the Day 3 activity and crayons. Then introduce the activity by reviewing:

*So far, we have divided shapes into two equal parts called **halves**. We can also divide shapes into **four** equal parts called **fourths**. Each equal part is called a **fourth**.*

- *Look at the pies in row 1. Which one is divided into four equal parts?* (the first one) *Color **one** part of the pie. This one part is called a **fourth**.*

- *Look at the crackers in row 2. Is the first one divided into four equal parts?* (no) *Is the second one divided into four equal parts?* (yes) *Color one-fourth of the cracker.*

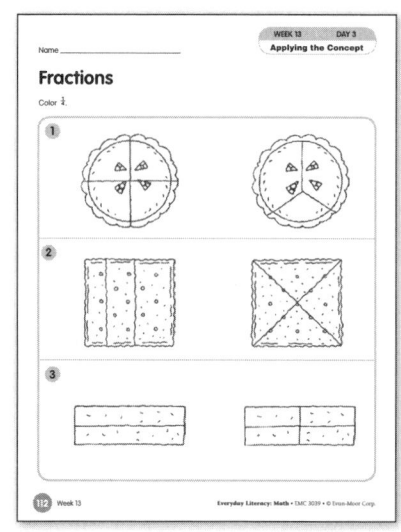

- *Look at the bar cookies in row 3. Is the first one divided into four equal parts?* (No; it is divided into halves.) *Is the second one divided into four equal parts?* (yes) *Color one-fourth of the cookie.*

Day 3 activity

Week 13

Everyday Literacy: Math • EMC 3039 • © Evan-Moor Corp.

Day 4 SKILLS

Geometry
- Understand that shapes can be divided into equal parts called halves or fourths
- Understand that decomposing a shape results in smaller parts

Literacy

Oral Language Development
- Respond orally to simple questions
- Use mathematical terms

Comprehension
- Make connections using illustrations, prior knowledge, or real-life experiences

Extending the Concept

Distribute the Day 4 activity and crayons. Then guide students through the activity by saying:

- *Point to number 1. Into how many equal parts is this shape divided?* (4) *Color one part.* (students color) *The first fraction says $\frac{1}{2}$ and the second one says $\frac{1}{4}$. What part did you color:* **one-half** *or* **one-fourth**? ($\frac{1}{4}$) *Circle the fraction $\frac{1}{4}$.*

- *Point to number 2. Into how many equal parts is this shape divided?* (2) *Color one part.* (students color) *What part did you color:* **one-half** *or* **one-fourth**? ($\frac{1}{2}$) *Circle the fraction $\frac{1}{2}$.*

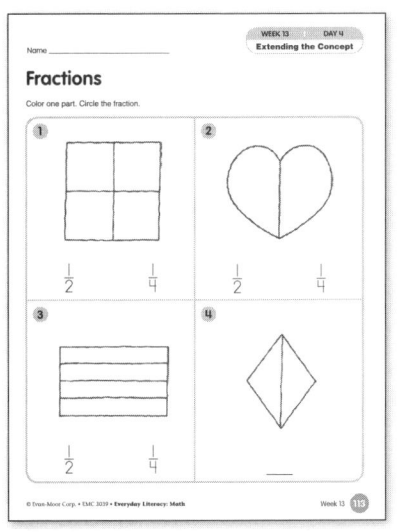

- *Point to number 3. Into how many equal parts is this shape divided?* (4) *Color one part.* (students color) *What part did you color:* **one-half** *or* **one-fourth**? ($\frac{1}{4}$) *Circle the fraction $\frac{1}{4}$.*

- *Point to number 4. Into how many equal parts is this shape divided?* (2) *Color one part.* (students color) *What part did you color:* **one-half** *or* **one-fourth**? ($\frac{1}{2}$) *Write the fraction $\frac{1}{2}$ on the line.*

Day 4 activity

Day 5 SKILLS

Geometry
- Understand that shapes can be divided into equal parts called halves or fourths

Mathematical Thinking and Reasoning
- Use math to solve problems

Home–School Connection p. 114
Spanish version available (see p. 2)

Hands-on Math Activity

Reinforce this week's math concept with the following hands-on activity:

Materials: construction paper circles the size of a pizza; decorations to be used as "toppings," such as dot stickers for pepperoni, small pompoms for sausage, paper hole reinforcements for olives, and shredded paper for cheese; glue

Activity: Divide students into small groups. Place the toppings in the center of the group for them to share. Distribute a paper pizza "crust" to each student. Have students fold the crust in halves or fourths and then open it so they can see the creases.

Suggest to students various pizza combinations, such as $\frac{1}{2}$ pepperoni and $\frac{1}{2}$ cheese; $\frac{1}{4}$ pepperoni, $\frac{1}{4}$ olive, $\frac{1}{4}$ sausage, and $\frac{1}{4}$ cheese. Then have them create their pizzas however they like, following the creases.

Have students describe their pizza creations to the group, remembering to use the fractions $\frac{1}{2}$ and $\frac{1}{4}$ as appropriate.

Name _____

Fractions

WEEK 13 | DAY 1
Confirming Understanding

Name _____

WEEK 13 | DAY 2
Reinforcing the Concept

Fractions

Listen. Color the circle next to **yes** or **no**.

1 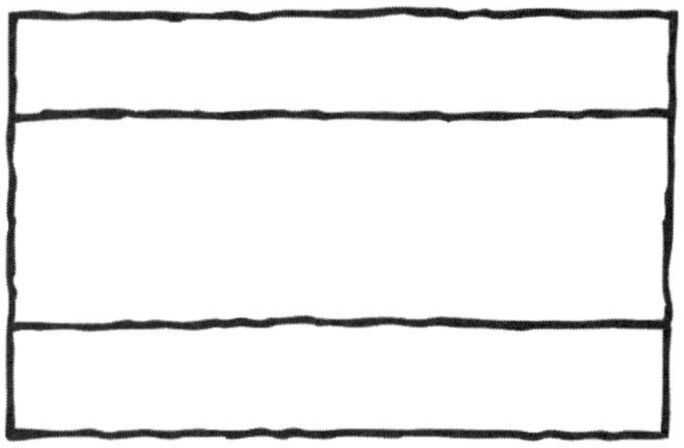
○ yes   ○ no

2 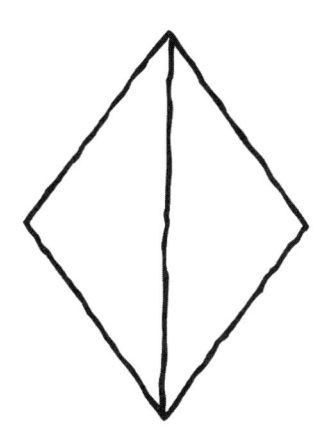
○ yes   ○ no

3 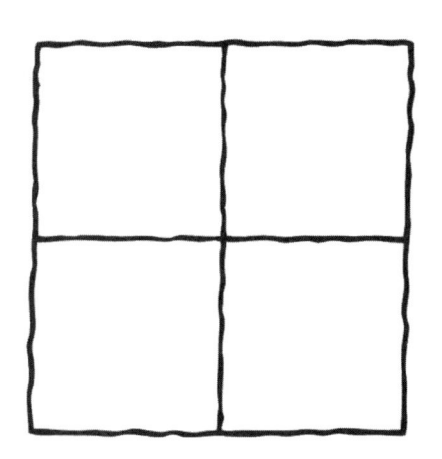
○ yes   ○ no

4 
○ yes   ○ no

© Evan-Moor Corp. • EMC 3039 • Everyday Literacy: Math   Week 13

Name _____

WEEK 13 | DAY 3
Applying the Concept

Fractions

Color $\frac{1}{4}$.

1

2

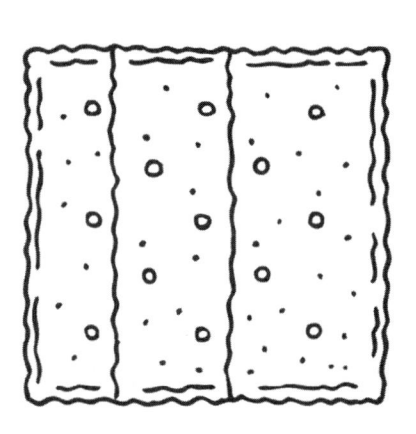

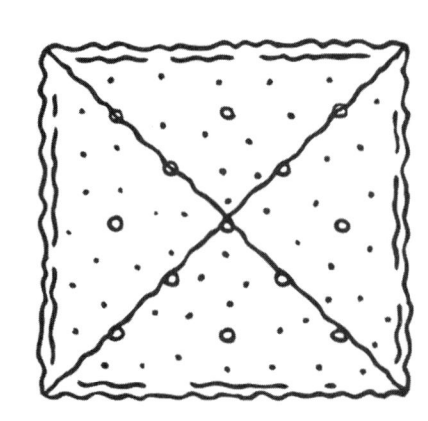

3

Week 13

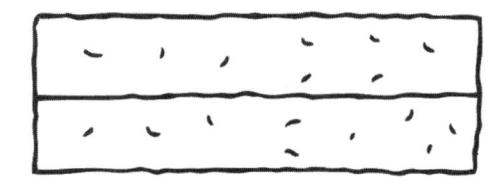

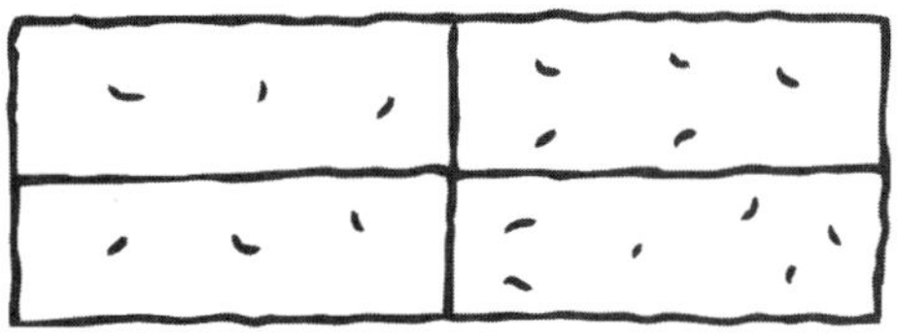

Everyday Literacy: Math • EMC 3039 • © Evan-Moor Corp.

Name _____

WEEK 13 | DAY 4
Extending the Concept

Fractions

Color one part. Circle the fraction.

1

$\frac{1}{2}$ $\frac{1}{4}$

2

$\frac{1}{2}$ $\frac{1}{4}$

3

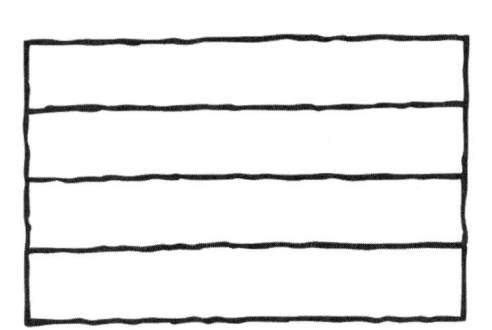

$\frac{1}{2}$ $\frac{1}{4}$

4

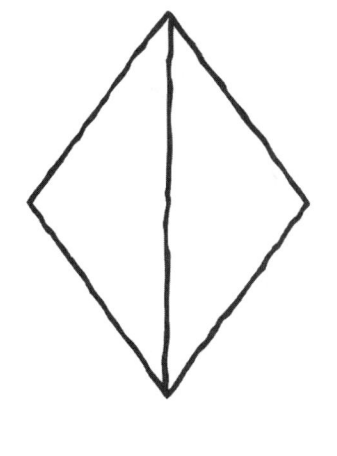

___

Name _____

What I Learned

What to Do
Have your child look at the picture below and tell you which pie is divided into halves. (the second one) Then have your child color each half a different color.

WEEK 13

Home–School Connection

Math Concept: Some shapes can be divided into halves and fourths.

To Parents
This week your child learned about fractions.

What to Do Next
Cut an orange into slices. Cut a slice into halves. Have your child explore how the two halves make a whole slice. Cut other slices into fourths; show your child that two fourths make a half.

114 Week 13

Everyday Literacy: Math • EMC 3039 • © Evan-Moor Corp.

WEEK 14

Concept
Measurement is used to determine the exact size of an object.

Ways to Measure

Math Objective:
To help students order and compare the length of objects and use nonstandard units to measure

Math Vocabulary:
length, longest, measure, order, shortest, units

Day 1 SKILLS

Measurement
- Use a nonstandard unit to measure

Literacy

Oral Language Development
- Respond orally to simple questions
- Use measurement terms

Comprehension
- Make connections using illustrations, prior knowledge, or real-life experiences
- Answer questions about key details in a text read aloud

Introducing the Concept

Model nonstandard measurement. Distribute a marker, new pencil, and crayon to each student. Lead them in comparing the length of the writing tools. Say:

- *First, lay the marker down. Put the pencil under the marker. Put the crayon at the bottom. Line up the tools so they start at the same point.*
- *Move the **longest** object to the top. Put the **shortest** one at the bottom. Name the objects from **shortest** to **longest**.* (crayon, marker, pencil) *Now say the order from **longest** to **shortest**.* (pencil, marker, crayon)

Distribute 10 large paper clips to each student. Say:

*We can use paper clips as **units**, or tools, to measure. Line up paper clips end to end under the pencil. Count the paper clips. How many paper clips long is the pencil?* (students respond) *How long is the crayon?* (students respond)

Listening to the Story

Distribute the Day 1 activity. Say: *Look at the picture and listen as I read a story about mice and toothpicks.*

Tom and Tessa had a tiny mouse house in the wall of a big house. One day, the owner of the big house dropped a box of toothpicks. Great! Tom and Tessa had been looking for a new sofa. The small box just might work. But would the box fit in their tiny house? "Let's measure the box with toothpicks," said Tessa. The mice lined up a few toothpicks, end to end, along the front of the box. Three toothpicks long! They laid the three toothpicks on the floor where the sofa would go. A perfect fit!

Confirming Understanding

Develop the math concept by asking questions about the story. Ask:

- *What did the mice use to measure the length of the toothpick box?* (toothpicks)
- *The mice used toothpicks, but you can also use paper clips to measure the toothpick box. How many paper clips long is the toothpick box?* ($2\frac{1}{2}$)

Day 1 picture

© Evan-Moor Corp. • EMC 3039 • **Everyday Literacy: Math**

Week 14

Day 2 SKILLS

Measurement
- Use a nonstandard unit to measure

Literacy

Oral Language Development
- Respond orally to simple questions
- Use measurement terms

Comprehension
- Make connections using illustrations, prior knowledge, or real-life experiences
- Answer questions about key details in a text read aloud

Reinforcing the Concept

Reread the Day 1 story. Then reinforce this week's math concept by discussing the story. Say:

In our story, how did the mice measure the toothpick box? (They lined up toothpicks.)

Distribute the Day 2 activity, large paper clips, and crayons. Say:

- *We are going to use paper clips to measure the lamps. Point to the first lamp. Place a paper clip at its base, over the outline. Place another paper clip above that one. Place one last paper clip above that one. How many paper clips tall is the first lamp?* (3) *Write the number **3** on the line below this lamp.*

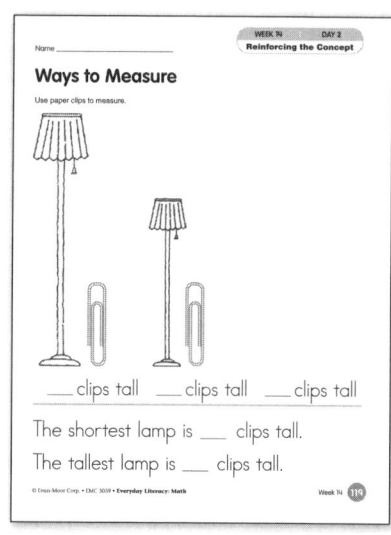

- *Point to the second lamp. Place a paper clip at its base. Now place another paper clip above that one. How many paper clips tall is the second lamp?* (2) *Write the number **2** on the line below this lamp.*

- *Now draw a lamp in the last space. Measure it with paper clips. How many paper clips tall is your lamp? Write that number on the line.*

- *How many paper clips tall is the shortest lamp? Write that number in the first blank.*

- *How many paper clips tall is the tallest lamp? Write that number in the second blank.*

Day 2 activity

Day 3 SKILLS

Measurement
- Use a nonstandard unit to measure

Literacy

Oral Language Development
- Respond orally to simple questions
- Use measurement terms

Comprehension
- Make connections using illustrations, prior knowledge, or real-life experiences

Applying the Concept

Distribute the Day 3 activity, large paper clips, and crayons. Then guide students through the activity by saying:

- *Look at the picture of the feather. Guess how many paper clips long it is. Write your guess in the box labeled **guess**.*

- *Now measure the feather with your paper clips. How many paper clips long is it?* (2) *Write the number **2** in the box labeled **measure**. Was your guess correct?* (Answers will vary.)

- *Look at the picture of the paintbrush. Guess how many paper clips tall it is. Write your guess in the box labeled **guess**.*

- *Now measure the paintbrush with your paper clips. How many paper clips tall is it?* (3) *Write the number **3** in the box labeled **measure**. Was your guess correct?* (Answers will vary.)

Day 3 activity

Week 14

Day 4 SKILLS

Measurement
- Use a nonstandard unit to measure

Literacy

Oral Language Development
- Respond orally to simple questions
- Use measurement terms

Comprehension
- Make connections using illustrations, prior knowledge, or real-life experiences

Extending the Concept

Distribute the Day 4 activity, large paper clips, and crayons. Then introduce the activity by reminding students of the Day 1 story:

- *Here is a picture of the inside of the mouse house. They are cleaning up, getting ready to bring in the toothpick box sofa. Let's find items that are taller or longer than a paper clip.*

- *Look at the picture. Without actually measuring, guess which items are taller than a paper clip. Draw a line under those items.*

Allow students time to complete the activity. After they have finished, say:

- *Now look at the items you marked. Measure each one with your paper clips. Were your guesses correct?* (students respond)

- *Which items in the mouse house are taller or longer than a paper clip? Color those items.* (mop, lamp, feather duster)

Day 4 activity

Day 5 SKILLS

Measurement
- Use a nonstandard unit to measure

Mathematical Thinking and Reasoning
- Use number concepts for a meaningful purpose

Home–School Connection p. 122
Spanish version available (see p. 2)

Hands-on Math Activity

Reinforce this week's math concept with the following hands-on activity:

Materials: large paper clips

Activity: Tell students that they will go on a "measure" hunt. To begin, have students link together 5 paper clips. Then have them go around the room and measure things that might be 5 or fewer paper clips long. Encourage each student to find three things that measure 5 or fewer paper clips.

Have students record their findings on large sticky notes. Have them represent each item with a drawing or a word, along with the number of paper clips it measured. For example, students may draw a pencil and label it with the number **3** to represent a pencil that is 3 paper clips long.

After students have finished, tell them you will show the results on a class graph. Draw a grid with 5 rows and label the rows **1, 2, 3, 4,** and **5 clips**.

Start the data collection by asking, *What did you find that is 1 paper clip long?* In the first row, have students place their sticky notes that represent the things that measure 1 paper clip. Continue with the other rows.

Analyze the results by asking questions such as, *Are there more items that are 3 paper clips long or 5 paper clips long? Is a pencil longer than a crayon?*

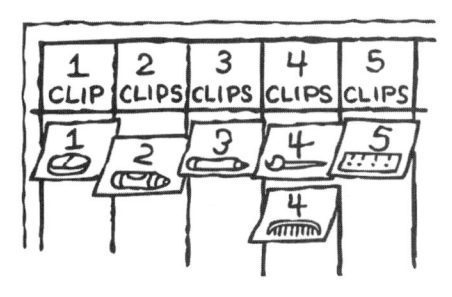

© Evan-Moor Corp. • EMC 3039 • **Everyday Literacy: Math**

Week 14

Name _____

Ways to Measure

Name _____

WEEK 14 | DAY 2
Reinforcing the Concept

Ways to Measure

Use paper clips to measure.

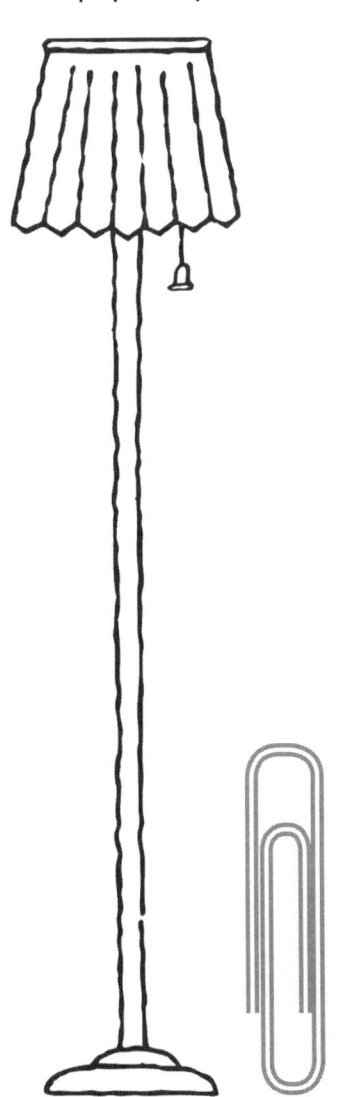

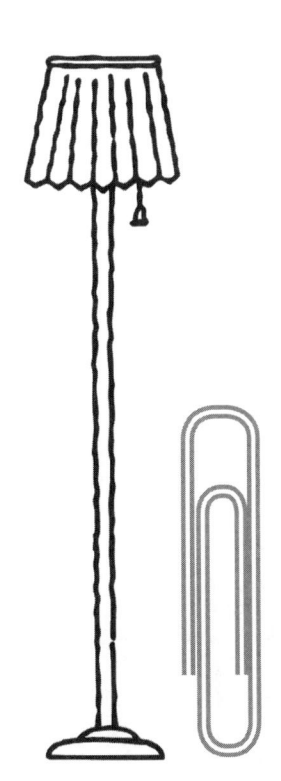

___ clips tall   ___ clips tall   ___ clips tall

The shortest lamp is ___ clips tall.

The tallest lamp is ___ clips tall.

Name _____

WEEK 14 | DAY 3
Applying the Concept

Ways to Measure

Guess. Then measure with a paper clip.

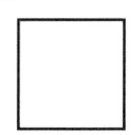

guess ☐   measure ☐

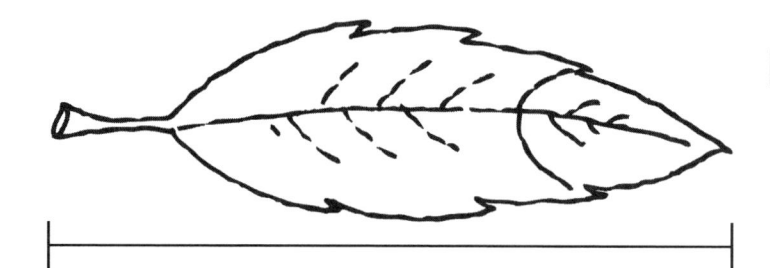

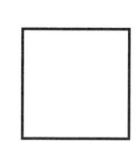

___

guess ☐   measure ☐

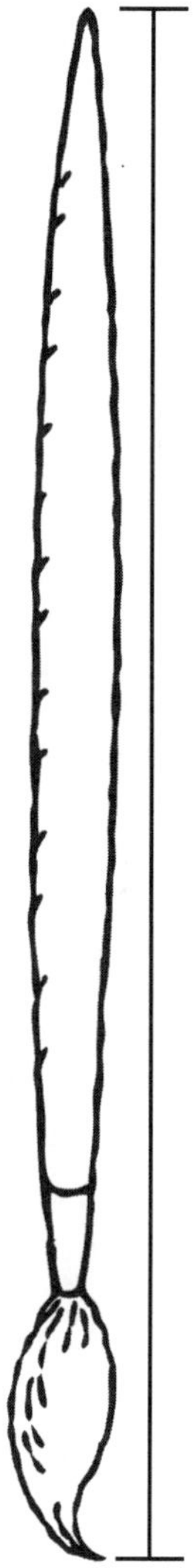

120 Week 14

Everyday Literacy: Math • EMC 3039 • © Evan-Moor Corp.

Name _____

WEEK 14 | DAY 4
Extending the Concept

Ways to Measure

Color the items that are taller than one paper clip.

Week 14

Name _____

What I Learned

What to Do
Have your child look at the picture below and use a toothpick or a birthday candle to measure different items. Then ask your child to compare two items. For example, ask, *What is longer, the feather or the paintbrush?*

WEEK 14

Home–School Connection

Math Concept: Measurement is used to determine the exact size of an object.

To Parents
This week your child learned to use nonstandard units of measure.

What to Do Next
Give your child a straw or some other tool to measure with. Challenge your child to find items in the house that are more or less the same length as the straw.

122 Week 14 Everyday Literacy: Math • EMC 3039 • © Evan-Moor Corp.

WEEK 15

Concept
Measurement is used to determine the exact size of objects.

Measure with Inches

Math Objective:
To help students use an inch ruler to measure

Math Vocabulary:
estimate, foot, height, inch, length, measure, ruler, vertically, width

Day 1 SKILLS

Measurement
- Use a ruler to measure

Literacy

Oral Language Development
- Respond orally to simple questions
- Use measurement terms

Comprehension
- Make connections using illustrations, prior knowledge, or real-life experiences
- Answer questions about key details in a text read aloud

Introducing the Concept

Distribute the Day 1 activity and rulers. Familiarize students with the marks on the ruler and what they stand for. Then say:

- *We can use an **inch ruler** to measure **length**, or how long something is. We can also measure **width**, or how wide something is. We can also measure **height**, or how tall something is. A **ruler** is one **foot** long. Look at your ruler. Tell me how many inches equal one foot.* (12)

- *Look at the activity page. Let's measure the tractor. Line up the end of your ruler with the end of the tractor. How long is the tractor?* (4 inches)

- *Now let's measure the row of cabbages. It is almost 6 inches long, so we will say it is **about** 6 inches. How long is the row?* (about 6 inches)

Listening to the Story

Redirect students' attention to the Day 1 activity. Say: *Look at the picture and listen as I read a story about Max, a boy who is measuring toys.*

> Max wanted to build a tiny farm. His farm had to fit in a shoe box. Max had a toy tractor and a plastic haystack. "I will measure these things to see if they will fit," thought Max. The tractor was **4 inches long**, and the haystack was **2 inches wide**. Max knew they would fit. He made the other pieces for his farm. He rolled balls of clay to look like heads of cabbage. He glued some grass on sticks to make a scarecrow, too. Max's tiny farm looked great!

Confirming Understanding

Distribute crayons. Develop the math concept by saying:

- *Use your ruler to measure the **length** of the tractor. How long is it?* (4 inches) *Measure the **width** of the haystack. Is the haystack **2 inches** wide?* (yes)

- *Measure the arms of the scarecrow to see how long they are.* (2 inches)

- *How long is the row of cabbages?* (about 6 inches) *Color **3 inches** of the cabbage row.* (3 heads of cabbage)

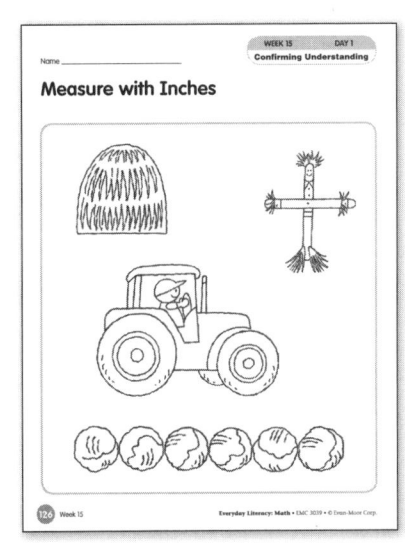

Day 1 picture

© Evan-Moor Corp. • EMC 3039 • Everyday Literacy: Math

Week 15

Day 2 SKILLS

Measurement
- Use a ruler to measure

Literacy

Oral Language Development
- Respond orally to simple questions
- Use measurement terms

Comprehension
- Make connections using illustrations, prior knowledge, or real-life experiences

Reinforcing the Concept

Reread the Day 1 story. Then reinforce this week's math concept by discussing the story. Say:

*To measure objects from our story, we used a ruler. Now let's guess, or **estimate**, the measurements of some vegetables.*

Distribute the Day 2 activity, crayons, and rulers. Then say:

- *Point to the first picture. It shows a corncob. Make a good guess, or **estimate**, of how many inches long it is. Do you think it is **1**, **2**, or **3** inches long? Write the number on the line after the word **estimate**.*

- *Next, use the ruler to **measure** the corncob to see if your estimate was correct. How many inches long is the corncob? (3) Write **3** on the line after the word **measure**.*

- *Point to the next picture. It shows a pea pod. First, **estimate** how many inches long it might be. Do you think it is **1**, **2**, or **3** inches long? Write the number on the line after the word **estimate**.*

- *Next, use the ruler to **measure** the exact length of the pea pod. How many inches long is it? (2) Write **2** on the line after the word **measure**.*

Repeat the process with the broccoli, having students estimate and measure how long it is.

Day 2 activity

Day 3 SKILLS

Measurement
- Use a ruler to measure

Literacy

Oral Language Development
- Respond orally to simple questions
- Use measurement terms

Comprehension
- Make connections using illustrations, prior knowledge, or real-life experiences

Applying the Concept

Distribute the Day 3 activity, rulers, and crayons. Then guide students through the activity by saying:

- *Look at the picture of a farm. First we will **estimate** and then we will **measure** with a ruler. Write your answers on the chart.*

- *Point to the barn. Estimate the **height**, or how tall it is. Write your estimate on the chart, in the row labeled "estimate."*

- *Now measure the barn. How tall is it? (2 inches) Write the number **2** on the chart, in the row labeled "measure."*

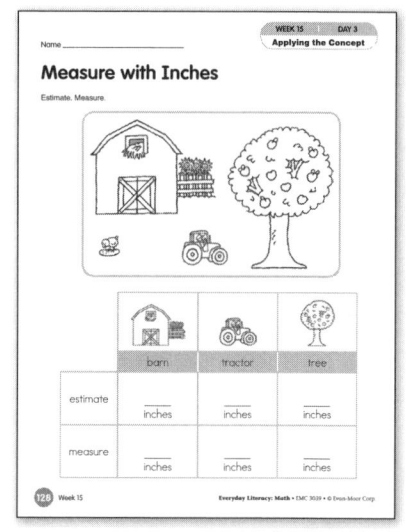

- *Point to the tractor. Estimate the **length**, or how long it is. Write your estimate on the chart, in the row labeled "estimate."*

- *Now measure the tractor. How long is it? (1 inch) Write the number **1** on the chart, in the row labeled "measure."*

- *Now, point to the tree. Estimate the **height**. Write your estimate on the chart, in the row labeled "estimate."*

- *Now measure the tree. How tall is it? (3 inches) Write the number **3** on the chart, in the row labeled "measure."*

Day 3 activity

Week 15

Everyday Literacy: Math • EMC 3039 • © Evan-Moor Corp.

Day 4
SKILLS

Measurement
- Use a ruler to measure

Literacy

Oral Language Development
- Respond orally to simple questions
- Use measurement terms

Comprehension
- Make connections using illustrations, prior knowledge, or real-life experiences

Extending the Concept

Distribute the Day 4 activity, rulers, and crayons. Model how to position the ruler vertically to measure height. Explain the word **vertically**. Then guide students through the activity by saying:

- *Look at the picture. We are going to measure the height of different things. Hold your ruler **vertically**, or up and down, on the page.*

- *First, measure the chicken coop door. Line up the end of your ruler with the bottom of the door. How tall is it?* (2 inches) *Write the number **2** in the box.*

- *Next, measure the post under the chicken coop. Line up the end of your ruler with the bottom of the post. How tall is the post?* (1 inch) *Write the number **1** in the box.*

- *Next, measure the height of the whole chicken coop. Line up the end of your ruler with the bottom of the chicken coop. How tall is the coop?* (4 inches) *Write the number **4** in the box.*

- *Now measure the height of the tree. Line up the end of your ruler with the bottom of the tree. How tall is it?* (5 inches) *Write the number **5** in the box.*

Day 4 activity

Day 5
SKILLS

Measurement
- Use a ruler to measure

Mathematical Thinking and Reasoning
- Record observations and data with pictures and other symbols

Home–School Connection p. 130
Spanish version available (see p. 2)

Hands-on Math Activity

Reinforce this week's math concept with the following hands-on activity:

Materials: rulers, sticky notes

Preparation: Draw a class chart with rows labeled **1 inch** through **12 inches**.

Activity: Divide students into groups (or pairs). Have each group be responsible for finding one item that is anywhere from 1 inch to 12 inches in length. For example, Group 1 could be responsible for finding something that measures 1 inch. Group 2 could be responsible for finding something that measures 2 inches, and so on.

After finding an item with the designated measurement, each group should write the item's name or draw its picture on a sticky note.

After students have finished, invite them to populate the chart. Begin by asking, *What item measures 1 inch?* (students respond) Have students place their illustrated sticky note in the row labeled **1 inch**. Continue with the rest of the measurements.

© Evan-Moor Corp. • EMC 3039 • **Everyday Literacy: Math**   Week 15

Name _____

WEEK 15 | DAY 1
Confirming Understanding

Measure with Inches

126 Week 15

Everyday Literacy: Math • EMC 3039 • © Evan-Moor Corp.

Name _____

WEEK 15 | DAY 2
Reinforcing the Concept

Measure with Inches

Estimate. Measure.

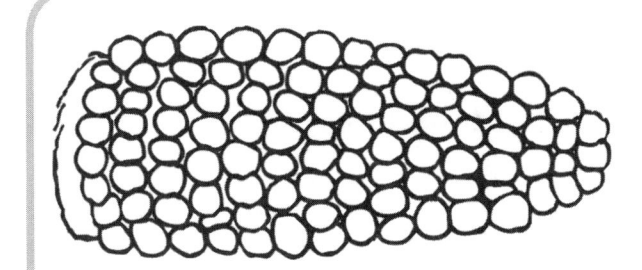

estimate: ___ inches

measure: ___ inches

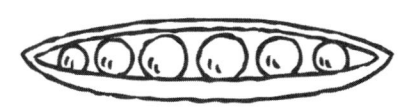

estimate: ___ inches

measure: ___ inches

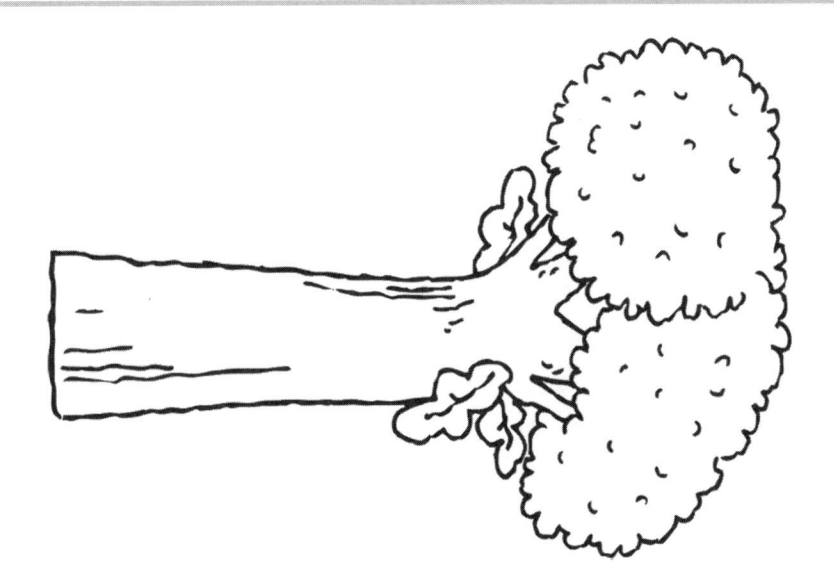

estimate: ___ inches

measure: ___ inches

© Evan-Moor Corp. • EMC 3039 • Everyday Literacy: Math   Week 15

Name _____

WEEK 15 | DAY 3
Applying the Concept

Measure with Inches

Estimate. Measure.

	barn	tractor	tree
estimate	_____ inches	_____ inches	_____ inches
measure	_____ inches	_____ inches	_____ inches

Name _____

WEEK 15 | DAY 4
Extending the Concept

Measure with Inches

Measure.

____ inches

____ inches

____ inches

____ inches

Week 15

Name _____

What I Learned

What to Do
Have your child look at the picture below. Help your child use a ruler to measure the items indicated by the lines. Then ask your child to measure other items, such as the ramp, the roof, and the chicken. Then have your child color the picture.

WEEK 15

Home–School Connection

Math Concept: Measurement is used to determine the exact size of objects.

To Parents
This week your child learned to measure using inches.

____ inches

____ inches

____ inches

____ inches

What to Do Next
Help your child identify things in your house that might measure 12 or fewer inches, and then estimate their length. Then ask your child to verify by measuring them with a ruler.

130 Week 15

Everyday Literacy: Math • EMC 3039 • © Evan-Moor Corp.

WEEK 16

Concept
Time can be measured.

Time for Fun!

Math Objective:
To help students tell and write time in hours and half-hours using analog and digital clocks

Math Vocabulary:
clock, digital clock, hour, hour hand, minute, minute hand, o'clock, time, watch

Day 1 SKILLS

Measurement
- Tell time to the hour and half-hour

Literacy

Oral Language Development
- Respond orally to simple questions
- Use vocabulary related to time concepts

Comprehension
- Make connections using illustrations, prior knowledge, or real-life experiences
- Answer questions about key details in a text read aloud

Introducing the Concept

Have on hand an example of an analog clock and a digital clock both set to 7:00. Point to the analog clock and its two hands and say:

How is the hour hand different from the minute hand? (The hour hand is shorter.) *You say the word **o'clock** when you see the minute hand on the 12. What time is it?* (7 o'clock) *We write the time this way: 7:00.*

Show the digital clock and identify it as such. Then say:

- *How is a digital clock different from the other clock?* (no hands; shows only numbers) *For a digital clock, you say **o'clock** when you see 00 after the dots. What time is it?* (7 o'clock) *We write the time this way: 7:00.*
- *These are two kinds of clocks. But we write 7:00* (point to the written example) *and say 7 o'clock for each clock.*

Listening to the Story

Distribute the Day 1 activity. Say: *Look at the picture and listen as I read a story about two kinds of watches.*

Eva and Ethan are celebrating their birthday. They are turning seven. "Let's open our gifts on the count of 3," says Eva. "1, 2, 3, rip!" Wrapping paper flies through the air. The twins lift lids on small boxes. They each get a watch! "My watch has hands; does yours?" Ethan asks. "No, my watch only has numbers," says Eva. She looks at Ethan's watch. "But they both say it's **4:00**. Time for cake!"

Confirming Understanding

Distribute pencils. Develop the math concept by asking questions about the story. Ask:

- *How is a **watch** the same as a **clock**?* (Both tell time.) *How is a watch different from a clock?* (You wear a watch.)
- *Who has a **digital** watch, Eva or Ethan?* (Eva) *Circle the digital watch.*
- *The clock in the box shows what time the twins went to bed. Write the time on the other clock.* (9:00) *Read the time out loud.* (9 o'clock)

Day 1 picture

© Evan-Moor Corp. • EMC 3039 • Everyday Literacy: Math

Week 16 131

Day 2
SKILLS

Measurement
- Tell time to the hour and half-hour

Literacy

Oral Language Development
- Respond orally to simple questions
- Use vocabulary related to time concepts

Comprehension
- Make connections using illustrations, prior knowledge, or real-life experiences

Reinforcing the Concept

Reread the Day 1 story. Then reinforce this week's math concept by discussing the story. Say:

In our story we saw two ways to tell time. One watch had hands and one had only numbers.

Distribute the Day 2 activity and pencils. Say:

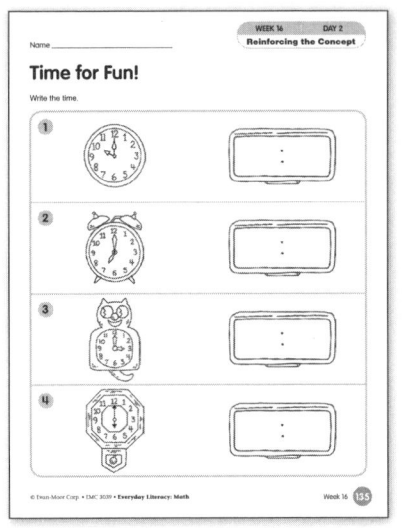

- *Point to the first clock. What time is it? (10:00) Yes, the hour hand is pointing to the 10. The minute hand is pointing to the 12. Write 10:00 on the digital clock. Write the number 10, then write 00 after the dots.*

- *Point to the second clock. What time is it? (7:00) Yes, the hour hand is pointing to the 7. The minute hand is pointing to the 12. Write 7:00 on the digital clock. What is happening where you live around 7:00 in the morning? (students respond)*

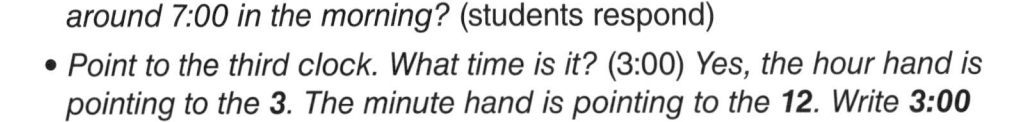

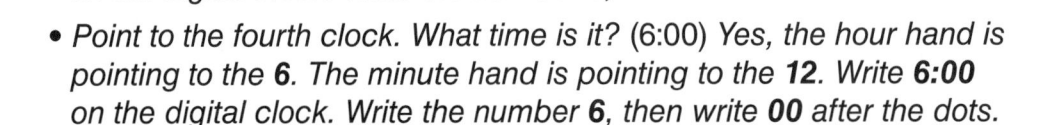

- *Point to the third clock. What time is it? (3:00) Yes, the hour hand is pointing to the 3. The minute hand is pointing to the 12. Write 3:00 on the digital clock. Write the number 3, then write 00 after the dots.*

- *Point to the fourth clock. What time is it? (6:00) Yes, the hour hand is pointing to the 6. The minute hand is pointing to the 12. Write 6:00 on the digital clock. Write the number 6, then write 00 after the dots.*

Day 2 activity

Day 3
SKILLS

Measurement
- Tell time to the hour and half-hour

Literacy

Oral Language Development
- Respond orally to simple questions
- Use vocabulary related to time concepts

Comprehension
- Make connections using illustrations, prior knowledge, or real-life experiences

Applying the Concept

Distribute the Day 3 activity and pencils. Then guide students through the activity by saying:

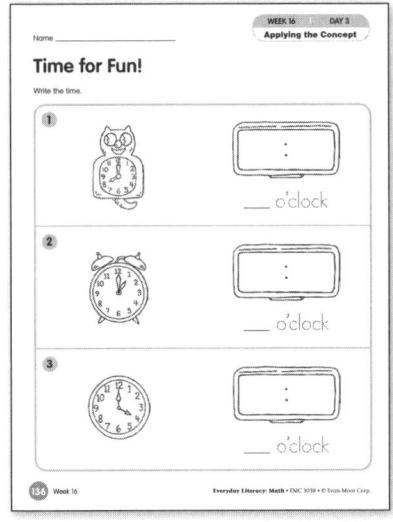

- *Point to the first clock. What time is it? (8:00) Yes, the hour hand is pointing to the 8. Write 8:00 on the digital clock. Then write 8 on the line and trace the word o'clock. Read the time again: 8 o'clock.*

- *Point to the second clock. What time is it? (1:00) Yes, the hour hand is pointing to the 1. Write 1:00 on the digital clock. Then write 1 on the line and trace the word o'clock. Read the time again: 1 o'clock.*

- *Point to the third clock. What time is it? (4:00) Yes, the hour hand is pointing to the 4. Write 4:00 on the digital clock. Then write 4 on the line and trace the word o'clock. Read the time again: 4 o'clock.*

Day 3 activity

Week 16

Everyday Literacy: Math • EMC 3039 • © Evan-Moor Corp.

Day 4
SKILLS

Measurement
- Tell time to the hour and half-hour

Literacy

Oral Language Development
- Respond orally to simple questions
- Use vocabulary related to time concepts

Comprehension
- Make connections using illustrations, prior knowledge, or real-life experiences

Extending the Concept

Use an analog clock to introduce the concept of telling time to the half-hour. Say:

*We learned that when the minute hand is on the 12, we use the word **o'clock** to tell time. In one **hour**, the minute hand goes all the way around the clock and lands on the 12.*

*In a **half-hour**, the minute hand goes halfway around the clock, to the **6**. So when the time is 2:30, the hour hand is halfway between the **2** and the **3** and the minute hand is on the **6**. Say the time: **2:30**.*

Distribute the Day 4 activity and pencils.

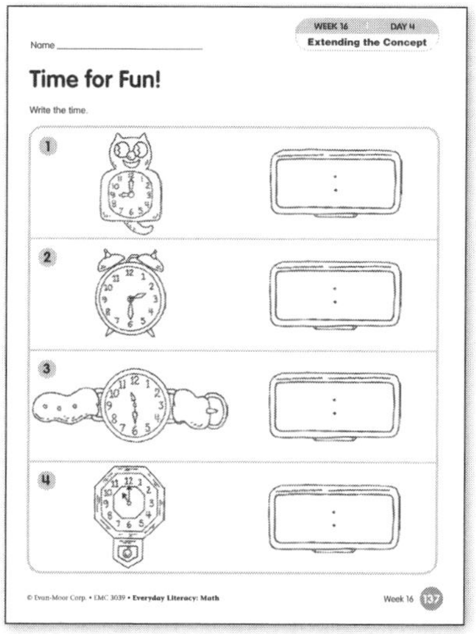

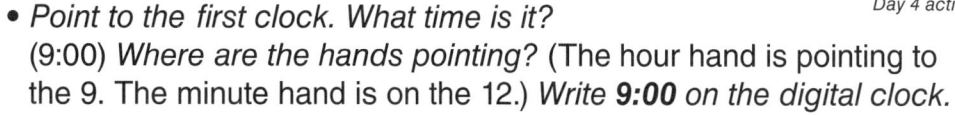

- *Point to the first clock. What time is it?* (9:00) *Where are the hands pointing?* (The hour hand is pointing to the 9. The minute hand is on the 12.) *Write **9:00** on the digital clock.*

- *Point to the second clock. What time is it?* (2:30) *Where are the hands pointing?* (The hour hand is between the 2 and the 3. The minute hand is on the 6.) *Write **2:30** on the digital clock.*

- *Point to the watch. What time is it?* (11:30) *Where are the hands pointing?* (The hour hand is between the 11 and the 12. The minute hand is on the 6.) *Write **11:30** on the digital clock.*

- *Point to the last clock. What time is it?* (11:00) *Where are the hands pointing?* (The hour hand is pointing to the 11 and the minute hand is on the 12.) *Write **11:00** on the digital clock.*

Day 4 activity

Day 5
SKILLS

Measurement
- Tell time to the hour and half-hour

Mathematical Thinking and Reasoning
- Use number concepts for a meaningful purpose

Home–School Connection p. 138
Spanish version available (see p. 2)

Hands-on Math Activity

Reinforce this week's math concept with the following hands-on activity:

Materials: yardstick and ruler; sidewalk chalk; various times to the hour and half-hour written on index cards (make two cards for each time)

Activity: You will need a large, open space. Draw a large clock face on the ground with chalk. Have students stand around the clock.

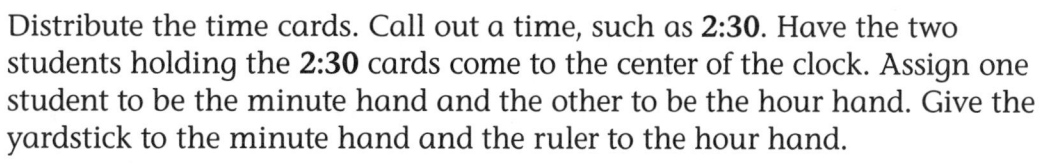

Distribute the time cards. Call out a time, such as **2:30**. Have the two students holding the **2:30** cards come to the center of the clock. Assign one student to be the minute hand and the other to be the hour hand. Give the yardstick to the minute hand and the ruler to the hour hand.

Have the two students show **2:30** by laying the ruler between the **2** and **3**, and laying the yardstick so it points to the **6**. Have the class decide if the positioning is correct.

Continue until all students have had a chance to show the time.

© Evan-Moor Corp. • EMC 3039 • Everyday Literacy: Math   Week 16

Name _____

Time for Fun!

WEEK 16 | DAY 1
Confirming Understanding

134 Week 16

Everyday Literacy: Math • EMC 3039 • © Evan-Moor Corp.

Name _____

WEEK 16 | DAY 2
Reinforcing the Concept

Time for Fun!

Write the time.

1.

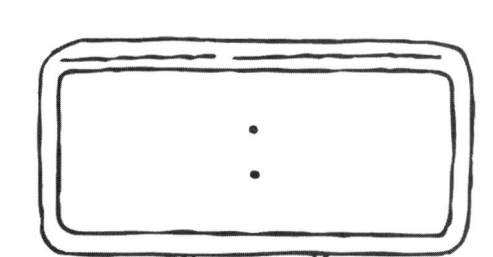

2.

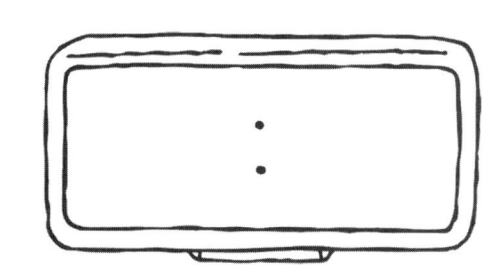

3.

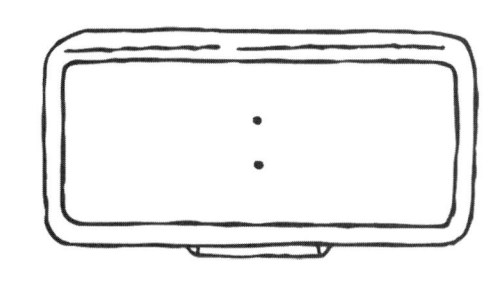

4.

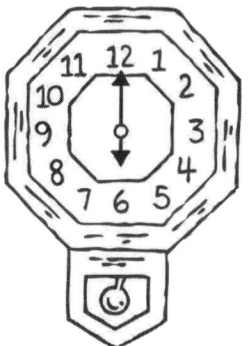

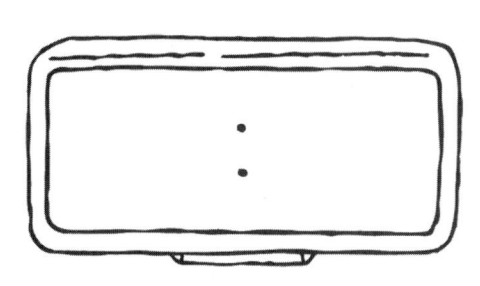

Week 16

Name _____

WEEK 16 | DAY 3
Applying the Concept

Time for Fun!

Write the time.

1. ___ o'clock

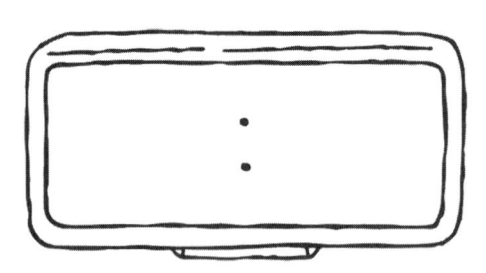

2. ___ o'clock

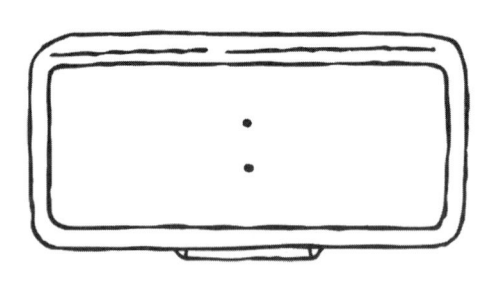

3. ___ o'clock

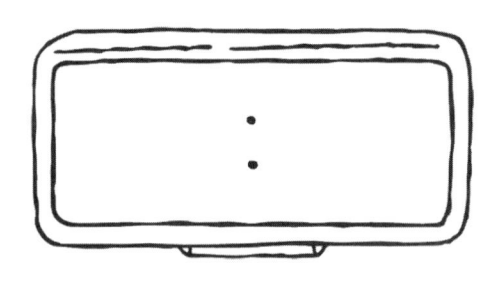

Name _____

WEEK 16 | DAY 4
Extending the Concept

Time for Fun!

Write the time.

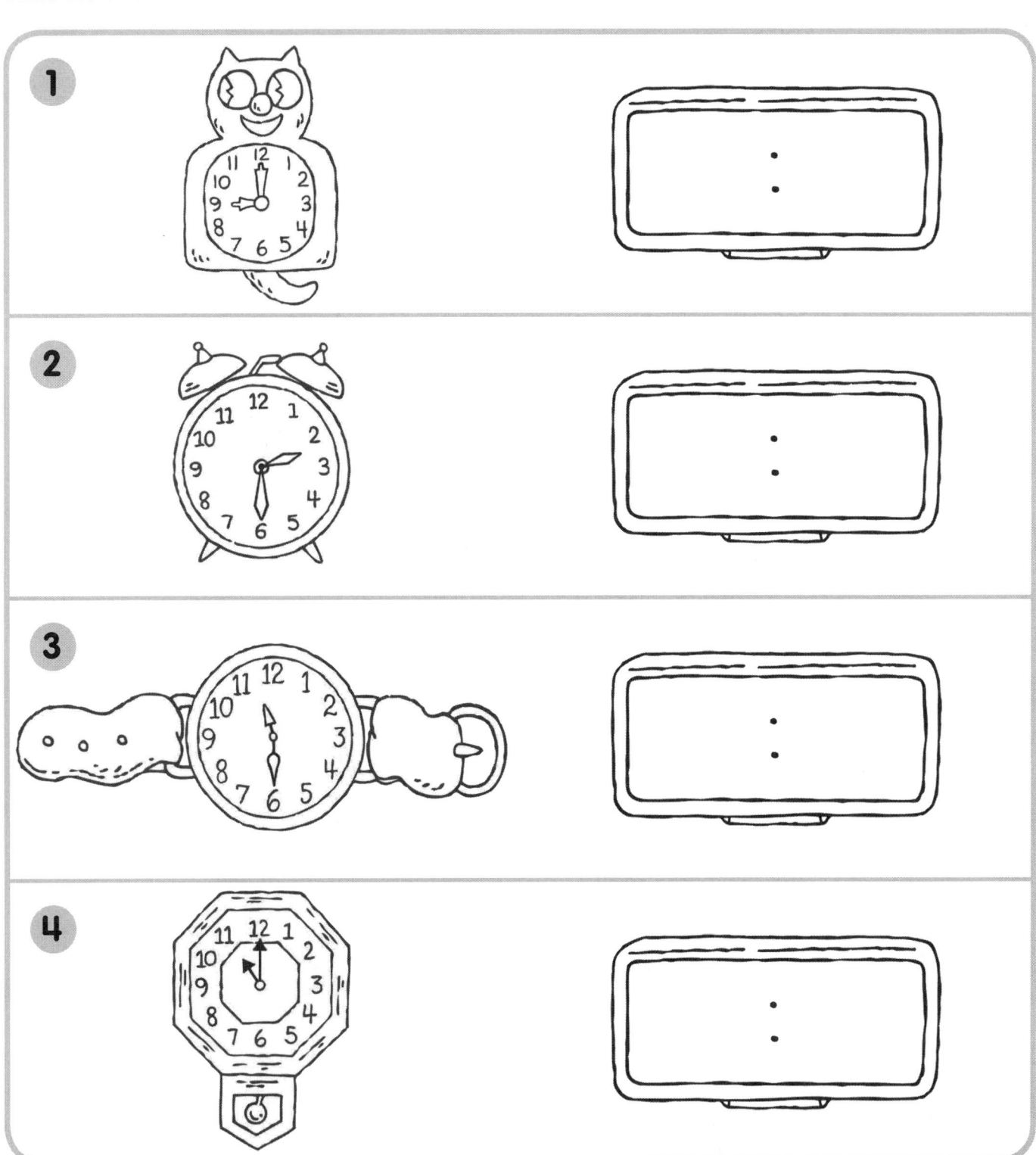

1.

2.

3.

4.

Week 16   137

Name _____

What I Learned

WEEK 16

Home–School Connection

What to Do
Have your child look at the picture below. Ask time-related questions, such as, *What time do the watches show? What time is shown on the clock in the box?* Then have your child write **9:00** on the blank digital clock.

Math Concept: Time can be measured.

To Parents
This week your child learned to tell time to the hour and half-hour.

What to Do Next
Make a paper plate clock, using a brad fastener to attach cardboard hands. Then have your child write the numbers on the clock and practice moving the hands to various times.

138 Week 16 — Everyday Literacy: Math • EMC 3039 • © Evan-Moor Corp.

WEEK 17

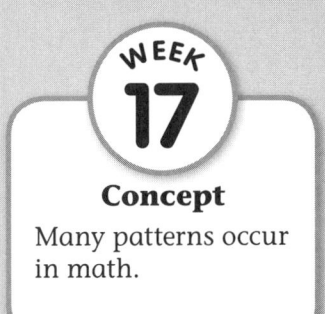

Concept
Many patterns occur in math.

Patterns

Math Objective:
To help students recognize, explain, and predict patterns

Math Vocabulary:
part, pattern, predict, repeat

Day 1 SKILLS

Algebra
- Recognize and form simple patterns

Literacy

Oral Language Development
- Respond orally to simple questions
- Use mathematical terms

Comprehension
- Make connections using illustrations, prior knowledge, or real-life experiences

Introducing the Concept

Before the lesson, draw a pattern of 3 colored hearts, all the same size: red, yellow, blue. Repeat the color pattern once and end with a red heart. Say:

- *Objects that repeat can form a pattern. Look at the color pattern in this drawing. Let's say the color pattern: **red, yellow, blue, red, yellow, blue, red**. What part of the pattern repeats?* (red, yellow, blue)

- *When you know the part that repeats, you can predict what will come next in the pattern. What color heart comes next?* (yellow)

Distribute paper and crayons. Have students work in pairs to create a different pattern using the colored hearts, such as **blue**, **red**, **yellow**. Invite students to say out loud the part of their pattern that repeats.

Listening to the Story

Distribute the Day 1 activity. Say: *Look at the picture and listen as I read a story about the patterns on a make-believe planet.*

Patrick drew a picture of a make-believe planet. He explained his drawing to his teacher. "This is Planet Pattern. You can see how the planet got its name. The rocks on this planet line up in a pattern of big, small, small, big, small, small. The spaceships and rockets stand in a pattern: spaceship, rocket, spaceship, rocket, spaceship, rocket. Planet Pattern is out of this world!"

Confirming Understanding

Distribute crayons. Develop the math concept by talking about the story. Say:

- *Say the pattern of the spaceships and rockets.* (spaceship, rocket, spaceship, rocket, spaceship, rocket) *Circle the part that repeats.* (spaceship, rocket)

- *Color the spaceships and rockets to make a color pattern. Use 2 colors.*

Day 1 picture

© Evan-Moor Corp. • EMC 3039 • **Everyday Literacy: Math**

Week 17

Day 2 SKILLS

Algebra
- Recognize and form simple patterns

Literacy

Oral Language Development
- Respond orally to simple questions
- Use mathematical terms

Comprehension
- Make connections using illustrations, prior knowledge, or real-life experiences

Reinforcing the Concept

Reread the Day 1 story. Then reinforce this week's math concept by guiding a discussion about the story. Say:

Our story featured rockets and spaceships in a pattern. What else was arranged in a pattern? (rocks)

Distribute the Day 2 activity and crayons. Say:

- *Point to row 1. Point to each item as we name it:* **spaceship**, **rocket**, **rocket**, **spaceship**, **rocket**, **rocket**. *What comes next,* **rocket** *or* **spaceship**? (spaceship) *Draw a circle around the spaceship.*

- *Point to row 2. Point to each item as we name it:* **star**, **star**, **moon**, **moon**, **star**, **star**. *What comes next,* **moon** *or* **star**? (moon) *Draw a circle around the moon.*

- *Point to row 3. Point to each item as we name it:* **sun**, **star**, **star**, **sun**, **star**, **star**. *What comes next,* **star** *or* **sun**? (sun) *Draw a circle around the sun.*

Day 2 activity

Day 3 SKILLS

Algebra
- Recognize and form simple patterns

Literacy

Oral Language Development
- Respond orally to simple questions
- Use mathematical terms

Comprehension
- Make connections using illustrations, prior knowledge, or real-life experiences

Applying the Concept

Distribute the Day 3 activity and pencils. Then guide students through the activity by saying:

Once you figure out the pattern, you can use letters to show that pattern.

- *Look at row 1. Point to each item as we name it:* **star**, **moon**, **star**, **moon**, **star**, **moon**. *What pattern repeats?* (star, moon) *On the lines, write an* **A** *for each star and a* **B** *for each moon. Start with* **A**.

- *What letters did you write for row 1?* (A, B, A, B, A, B) *What pattern repeats?* (A, B) *We can say that row 1 has an* **AB** *pattern.*

- *Look at row 2. Point to each item as we name it:* **dark moon**, **dark moon**, **light moon**, **dark moon**, **dark moon**, **light moon**. *What pattern repeats?* (dark, dark, light) *On the lines, write an* **A** *for each dark moon and a* **B** *for each light moon.*

- *What letters did you write for row 2?* (A, A, B, A, A, B) *What pattern repeats?* (A, A, B) *We can say that row 2 has an* **AAB** *pattern.*

- *Look at row 3. Point to each item as we name it:* **rocket**, **rocket**, **planet**, **planet**, **rocket**, **rocket**. *What pattern repeats?* (rocket, rocket, planet, planet) *On the lines, write an* **A** *for each rocket and a* **B** *for each planet.*

- *What letters did you write for row 3?* (A, A, B, B, A, A) *What pattern repeats?* (A, A, B, B) *We can say that row 3 has an* **AABB** *pattern.*

Day 3 activity

Day 4 SKILLS

Number Sense
- Count by 2s, 5s, or 10s

Algebra
- Recognize and form simple patterns

Literacy

Oral Language Development
- Respond orally to simple questions
- Use mathematical terms

Comprehension
- Make connections using illustrations, prior knowledge, or real-life experiences

Extending the Concept

Distribute the Day 4 activity and pencils. Then introduce the activity by saying:

We learned that objects and letters can form patterns. Numbers can also form patterns.

- *Look at row 1. Point to each number as we name it: **2, 4, 6, 8**. What two numbers come next?* (10, 12) *That's right. The pattern continues by adding 2 to the last number. In other words, we make a pattern by counting by twos. Write **10** and **12** on the lines to continue the pattern.*

- *Look at row 2. Point to each number as we name it: **10, 20, 30, 40**. What two numbers come next?* (50, 60) *Yes. The pattern continues by adding 10 to the last number. When we count by tens, we make a pattern. Write **50** and **60** on the lines to continue the pattern.*

- *Look at row 3. Point to each number as we name it: **5, 10**. What number comes next?* (15) *That's right. The pattern continues by adding 5 to the last number. Write the number **15** on the first line. Let's name the numbers so far: **5, 10, 15, 20, 25**. What's the next number?* (30) *Write **30** on the last line. We counted by fives.*

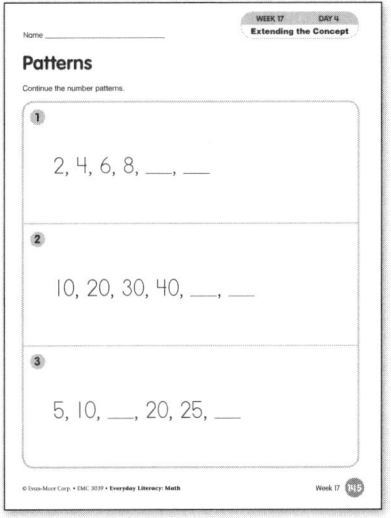

Day 4 activity

Patterns
Continue the number patterns.

1. 2, 4, 6, 8, ___, ___
2. 10, 20, 30, 40, ___, ___
3. 5, 10, ___, 20, 25, ___

Day 5 SKILLS

Algebra
- Recognize and form simple patterns

Mathematical Thinking and Reasoning
- Explore mathematical ideas through song or play

Home–School Connection p. 146
Spanish version available (see p. 2)

Circle Time Math Activity

Reinforce this week's math concept with the following circle time activity:

Preparation: Teach students two movements they will need to perform during the activity. One movement is called "the blast-off," and it is performed by extending the arms straight up toward the sky. Model the blast-off a few times.

The other movement is called "the moonwalk." It is performed by walking in place while lifting the knees in the air. Perform the moonwalk a few times.

Activity: Gather in a large outdoor space. Have students stand in a circle. Tell them they will form a pattern by using two movements. Select a student to start the pattern by performing the movement you dictate. Point to the first student and say, *Blast-off.* Point to the next student and say, *Blast-off.* Point to the next student and say, *Moonwalk.*

Extend the pattern with three more students: *blast-off, blast-off, moonwalk.* Then ask, *What comes next?* (blast-off) When students have figured out the pattern, have them continue around the circle on their own, until everyone is performing the appropriate movement.

Then have students create a different pattern.

© Evan-Moor Corp. • EMC 3039 • **Everyday Literacy: Math**   Week 17   141

Name _____

WEEK 17 | DAY 1
Confirming Understanding

Patterns

WEEK 17 | DAY 2
Reinforcing the Concept

Name _____

Patterns

Circle what comes next.

1

2

3

WEEK 17 | DAY 3
Applying the Concept

Name _____

Patterns

Show the pattern with A B.

1.
★ ☾ ★ ☾ ★ ☾
__ __ __ __ __ __

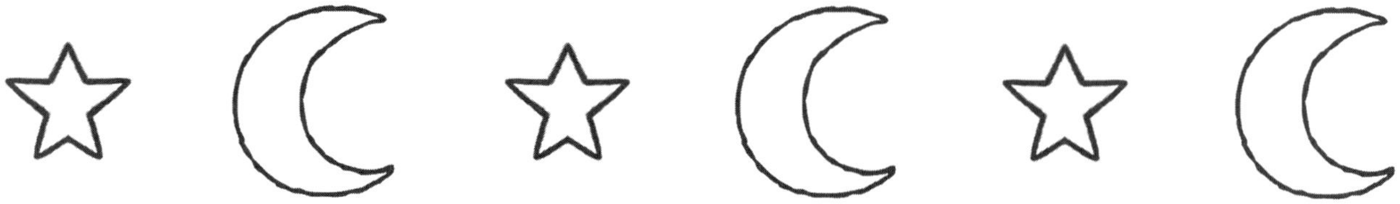

2.
☾ ☾ ☾ ☾ ☾ ☾
__ __ __ __ __ __

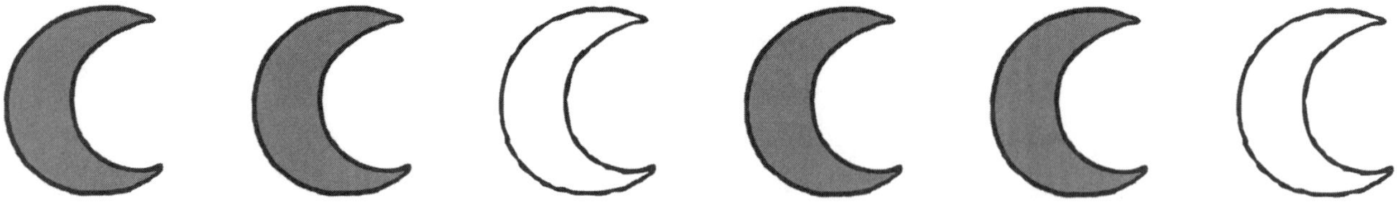

3.
🚀 🚀 🪐 🪐 🚀 🚀
__ __ __ __ __ __

144 Week 17   Everyday Literacy: Math • EMC 3039 • © Evan-Moor Corp.

Name _____

Patterns

Continue the number patterns.

1

2, 4, 6, 8, ___, ___

2

10, 20, 30, 40, ___, ___

3

5, 10, ___, 20, 25, ___

Name _____

What I Learned

What to Do
Have your child look at the picture below and identify the patterns. Then have your child color the picture, forming different patterns with colors.

WEEK 17

Home–School Connection

Math Concept: Many patterns occur in math.

To Parents
This week your child learned to recognize and form simple patterns.

What to Do Next
Encourage your child to find patterns throughout the house, such as on flooring, curtains, tiles, and dishes. Encourage your child to describe each pattern to you.

146 Week 17

Everyday Literacy: Math • EMC 3039 • © Evan-Moor Corp.

WEEK 18

Concept
Locations are indicated on coordinate grids.

Where Is It?

Math Objective:
To help students find locations on a grid and describe their spatial relationships

Math Vocabulary:
across, bottom, column, corner, grid, next to, row, up

Day 1 SKILLS

Geometry
- Find locations on a grid

Literacy

Oral Language Development
- Respond orally to simple questions
- Use mathematical terms

Comprehension
- Make connections using illustrations, prior knowledge, or real-life experiences
- Answer questions about key details in a text read aloud

Introducing the Concept

Distribute the Day 1 activity and identify it as a grid. Then say:

- *A **grid** is like a map divided into boxes. This grid shows where to find some of the animals in a zoo. Let's read the **rows**: **A**, **B**, **C**, **D**. Let's read the **columns**: **1**, **2**, **3**, **4**.*

- *To find a place on any grid, you start at the **bottom corner** on the left. Place your finger at the **start**.* (students respond) *Now let's find box **2B**. Move your finger from the start to the **2**. Then move your finger **up** to row **B**. What animals are at **2B**?* (lions)

- *Now let's tell **how** to find a place on the grid. How do we get to the elephants?* (Go right to 3; go up to C.) *Where are the elephants?* (3C)

Listening to the Story

Redirect students' attention to the Day 1 activity. Say: *Look at the grid and listen as I read a story about a girl and her dad reading a map at a zoo.*

Lila and her dad were at the zoo. Lila asked a worker where the giraffes were. The worker handed Lila a grid and said, "The giraffes are at **4D**." Lila moved her finger to find **4** on the grid. Then she moved her finger **up** to **D**. "The giraffes are far away," she said. "But the drinking fountains are at **4A**. We can stop there first." Dad looked at the grid. "My favorite animals are at **3D**. They're **next to** the giraffes."

Confirming Understanding

Distribute crayons. Develop the math concept by asking questions about the story. Ask:

- *Where are Dad's favorite animals on the grid?* (3D) *Find **3D** on the grid and circle it. What are Dad's favorite animals?* (zebras)

- *Where are the monkeys?* (1C) *Circle the **1C** box. Are the monkeys next to any other animals?* (no)

- *Draw an animal in **4B**.*

Ask volunteers to show their drawing and tell what animal they've drawn.

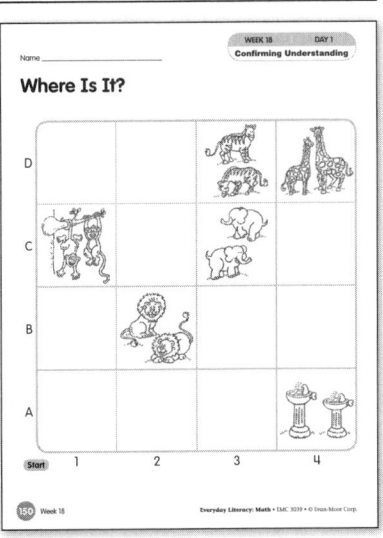

Day 1 picture

© Evan-Moor Corp. • EMC 3039 • **Everyday Literacy: Math**

Week 18   147

Day 2 SKILLS

Geometry
- Find locations on a grid

Literacy

Oral Language Development
- Respond orally to simple questions
- Use mathematical terms

Comprehension
- Make connections using illustrations, prior knowledge, or real-life experiences
- Answer questions about key details in a text read aloud

Reinforcing the Concept

Reread the Day 1 story. Then reinforce this week's math concept by guiding a discussion about the story. Say:

Our story was about a girl and her dad reading a grid at a zoo. Now here is a grid of a park.

Distribute the Day 2 activity and crayons. Say:

- *Let's find box **3C**. Put your finger at the start. Move across 3 boxes. Go up to C. What is it?* (a tree) *Draw a dot on the tree.*
- *Let's find box **4A**. Put your finger at the start. Move across 4 boxes. Then go up to A. What is it?* (a slide) *Circle the slide.*
- *Let's find box **2B**. Put your finger at the start. Move across 2 boxes. Then go up to B. What is it?* (flowers) *Draw a line under the flowers.*
- *Let's find box **3A**. Put your finger at the start. Move across 3 boxes. Then go up to A. What is it?* (swings) *Draw a box around the swings.*
- *Let's find box **5D**. Put your finger at the start. Move across 5 boxes. Then go up to D. What is it?* (a sun) *Color the sun yellow.*
- *Let's find box **1C**. Put your finger at the start. Move across to 1. Then go up to C. What is it?* (a butterfly) *Circle the butterfly.*

Day 2 activity

Day 3 SKILLS

Geometry
- Find locations on a grid

Literacy

Oral Language Development
- Respond orally to simple questions
- Use mathematical terms

Comprehension
- Make connections using illustrations, prior knowledge, or real-life experiences

Applying the Concept

Distribute the Day 3 activity and crayons. Then guide students through the activity by saying:

Now here is a grid with some shapes.

- *Let's find box **2B**. Put your finger at the start. Move across 2 boxes. Go up to B. Draw a star at **2B**.*
- *Where else do you see a star?* (1A)
- *Now find box **2D**. Put your finger at the start. Move across 2 boxes. Then go up to D. Draw a triangle at **2D**.*
- *Where else do you see a triangle?* (3B)
- *Let's find box **4C**. Put your finger at the start. Move across 4 boxes. Then go up to C. Draw a circle there.*
- *Where else do you see a circle?* (2C)
- *Now find box **1D**. Put your finger at the start. Move across to 1. Then go up to D. Draw a square there.*
- *Where else do you see a square?* (3D)

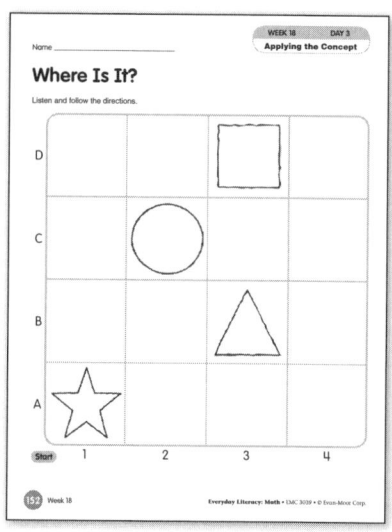

Day 3 activity

148 Week 18

Everyday Literacy: Math • EMC 3039 • © Evan-Moor Corp.

Day 4 SKILLS

Geometry
- Find locations on a grid

Literacy

Oral Language Development
- Use mathematical terms

Comprehension
- Make connections using illustrations, prior knowledge, or real-life experiences

Extending the Concept

Distribute the Day 4 activity and crayons. Guide students through the activity by saying:

We are going to color boxes on this grid. We will use red, blue, orange, and green. Listen carefully as I tell you which boxes to color.

- *First, find **1A** and color it red.*
- *Then find **1E** and color it red.*
- *After that, color **3C** red.*
- *Next, color **5A** red.*
- *Finally, color **5E** red.*
- *Now let's color the blue boxes: **1B, 1D, 2A, 2E, 4A, 4E, 5B, 5D**.*
- *Now let's color the orange boxes: **2B, 2D, 4B, 4D**.*
- *Last, let's color the green boxes: **1C, 2C, 3A, 3B, 3D, 3E, 4C, 5C**.*
- *Are all your boxes colored? Look at the pretty design you made!*

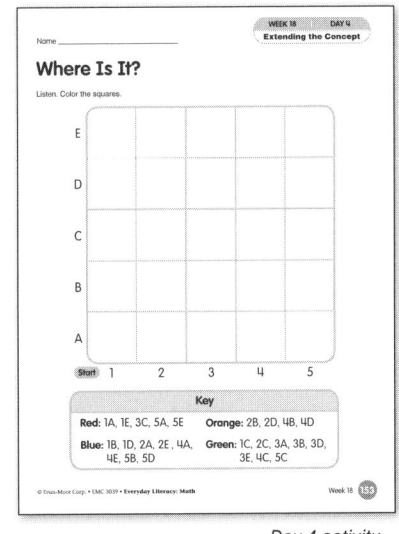

Day 4 activity

Day 5 SKILLS

Geometry
- Find locations on a grid

Mathematical Thinking and Reasoning
- Explore mathematical ideas through song or play

Home–School Connection p. 154
Spanish version available (see p. 2)

Hands-on Math Activity

Reinforce this week's math concept with the following hands-on activity:

Materials: one 3 x 3 construction paper grid for each pair of students, black beans, white beans, lima beans

Preparation: Label each grid with **1, 2, 3** and **A, B, C** as shown. Make game pieces by writing **1A, 1B, 1C, 2A, 2B, 2C, 3A, 3B, 3C** on lima beans. Each pair of students will get a grid and a set of game pieces.

Activity: Divide students into pairs: Partner A and Partner B. Give partners a grid to share. Give Partner A a few white beans. Give Partner B a few black beans. Then place the labeled lima beans in a container for partners to share.

To begin, have Partner A take a lima bean and read aloud the coordinate, for example, **2A**. Then have him or her place a white bean on the corresponding square on the grid.

Then have Partner B take a turn, marking his or her square with a black bean. The first to get three in a row is the winner. Alternatively, the first to get four corners can be the winner.

© Evan-Moor Corp. • EMC 3039 • **Everyday Literacy: Math**   Week 18

Name _____

WEEK 18 | DAY 1
Confirming Understanding

Where Is It?

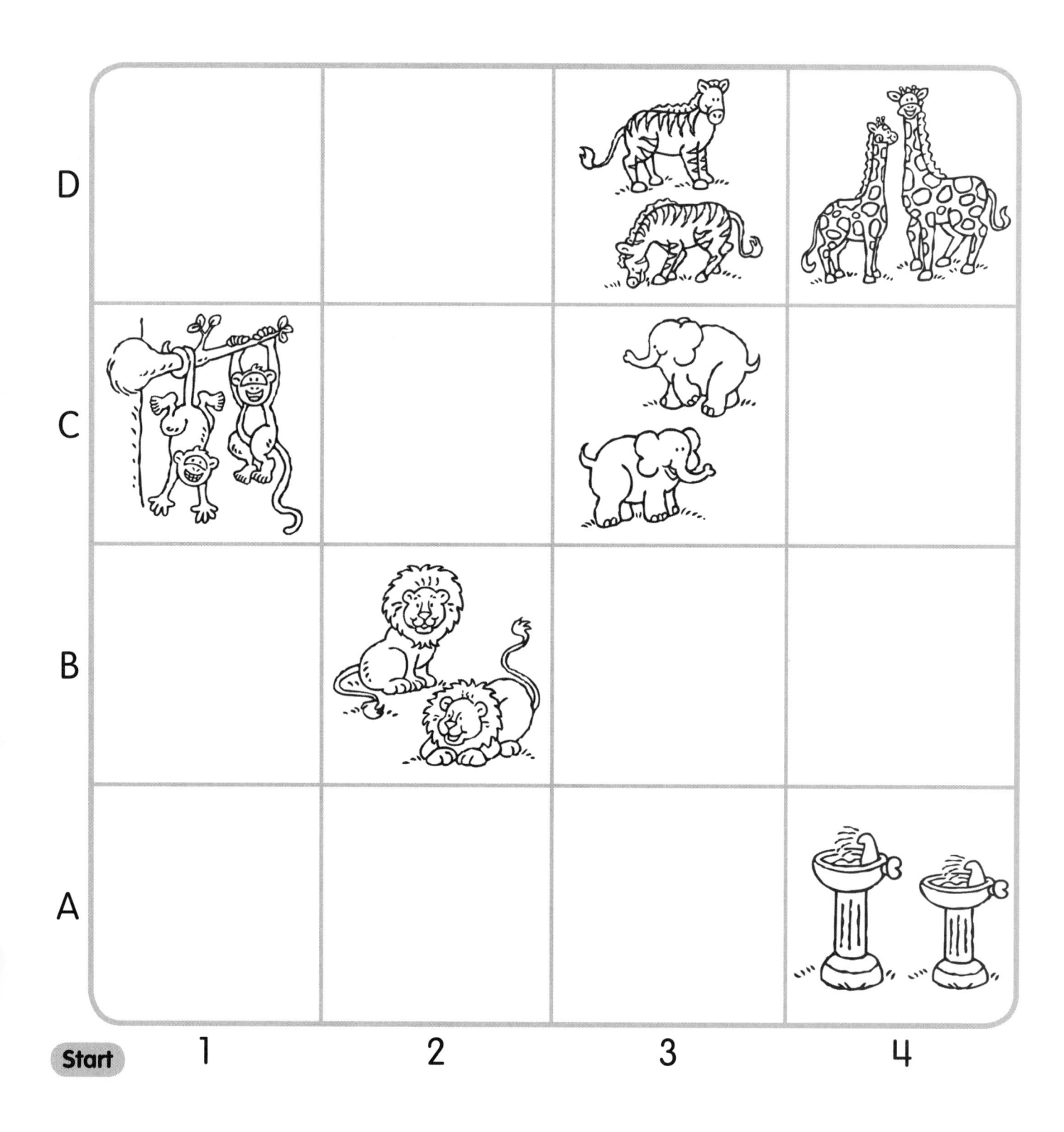

	1	2	3	4
D			zebras	giraffes
C	monkeys		elephants	
B		lions		
A				birdbaths

Start

150 Week 18

Everyday Literacy: Math • EMC 3039 • © Evan-Moor Corp.

Name _____

WEEK 18 | DAY 2
Reinforcing the Concept

Where Is It?

Listen and follow the directions.

Start 1 2 3 4 5

Week 18 151

© Evan-Moor Corp. • EMC 3039 • **Everyday Literacy: Math**

Name _____

WEEK 18 | DAY 3
Applying the Concept

Where Is It?

Listen and follow the directions.

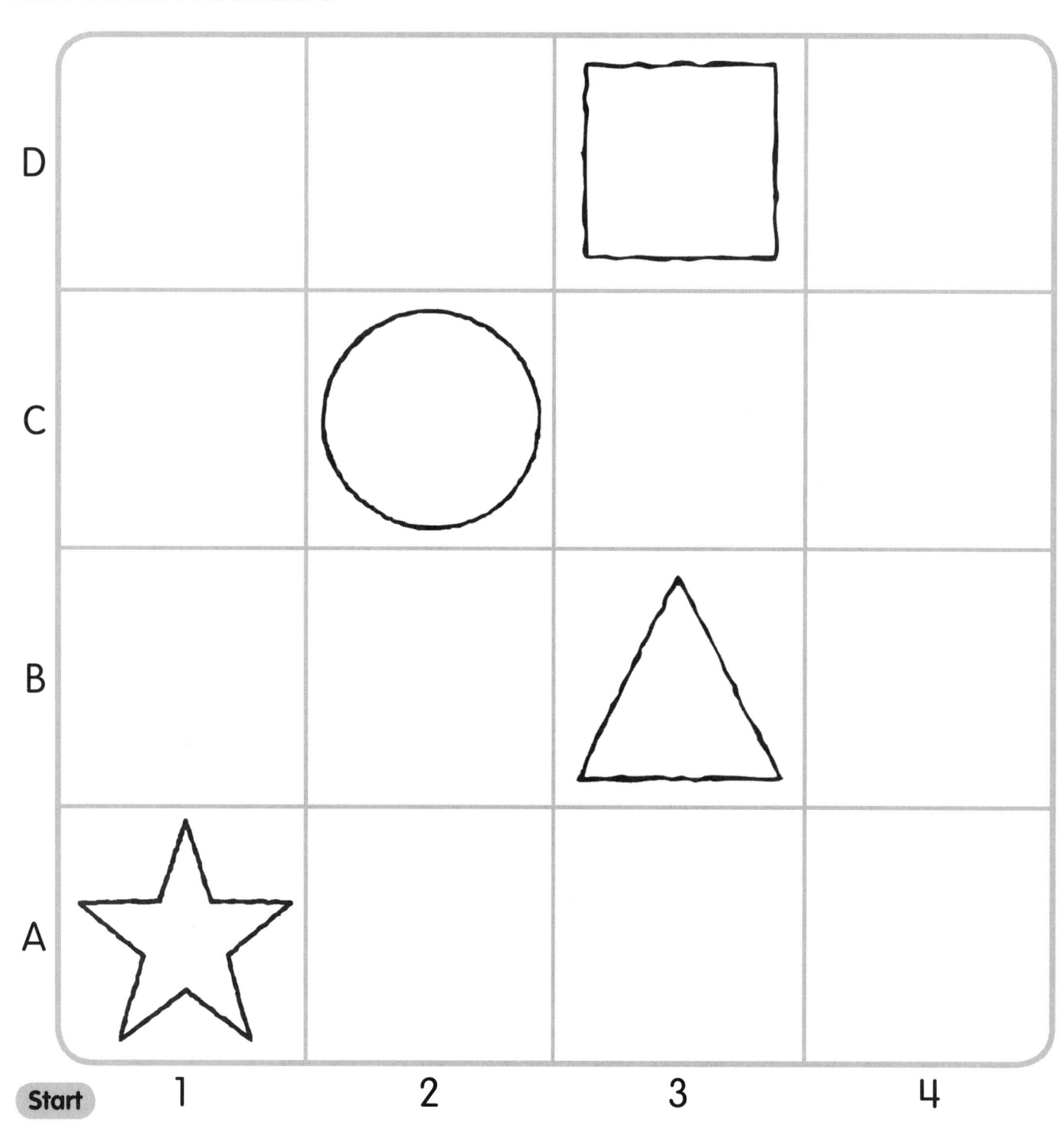

Start

Name _____

WEEK 18 | DAY 4
Extending the Concept

Where Is It?

Listen. Color the squares.

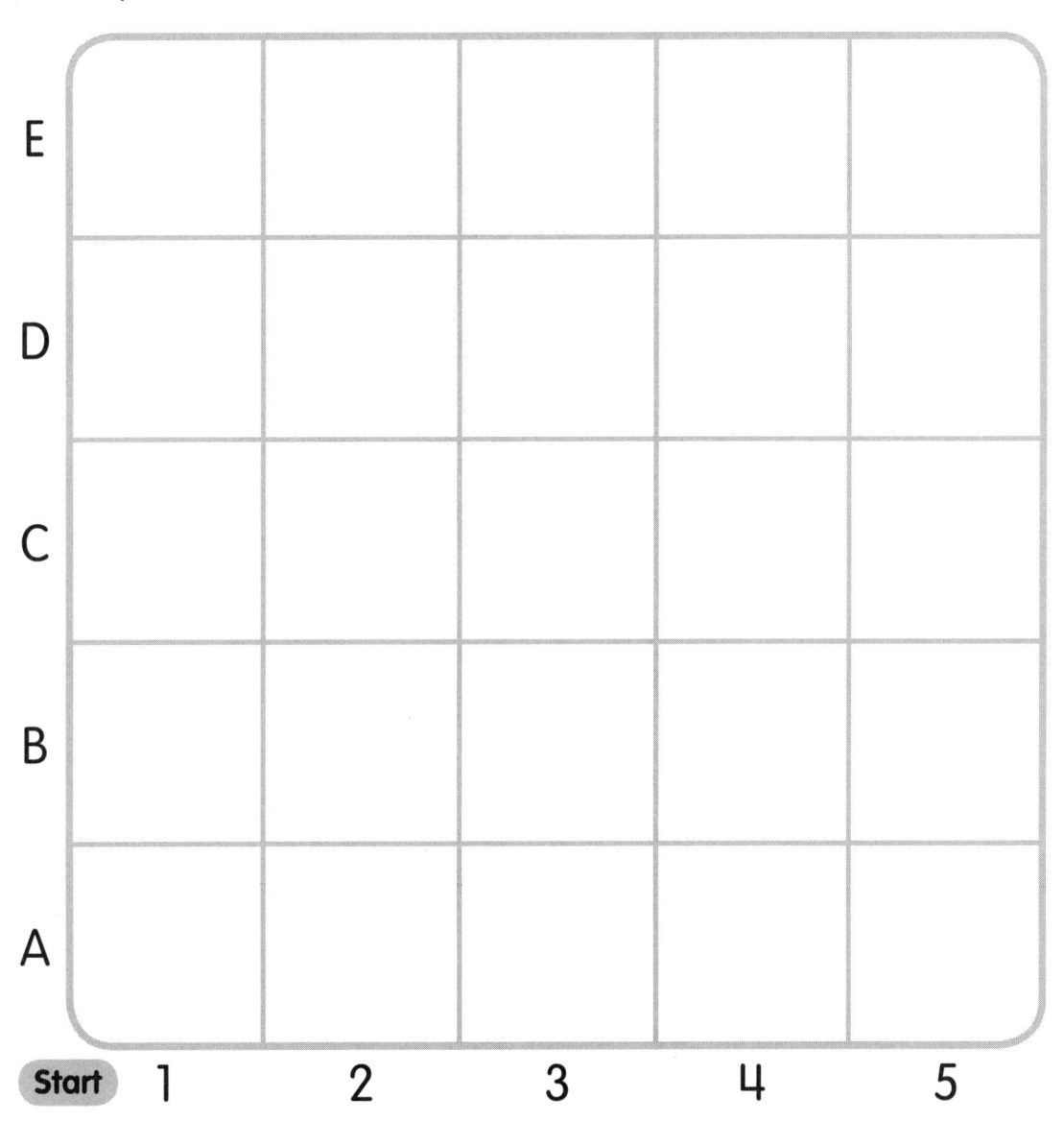

	1	2	3	4	5
E					
D					
C					
B					
A					

Start 1 2 3 4 5

Key

Red: 1A, 1E, 3C, 5A, 5E   **Orange:** 2B, 2D, 4B, 4D

Blue: 1B, 1D, 2A, 2E, 4A, 4E, 5B, 5D   **Green:** 1C, 2C, 3A, 3B, 3D, 3E, 4C, 5C

© Evan-Moor Corp. • EMC 3039 • *Everyday Literacy: Math*

Name _____

What I Learned

WEEK 18

Home–School Connection

What to Do
Have your child look at the grid below. Ask your child questions about animals on the grid, such as, *Where are the lions?* (at 2B) Then ask your child to draw two snakes at **1D**.

Math Concept: Locations are indicated on coordinate grids.

To Parents
This week your child learned to find locations on a grid.

	1	2	3	4
D				
C				
B				
A				

Start

What to Do Next
Use sidewalk chalk to draw a 4 x 4 grid on the pavement. Label it as above, with numbers and letters. Then call out place names, such as **3C**, **4D**, and have your child aim for those boxes with a pebble.

154 Week 18

Everyday Literacy: Math • EMC 3039 • © Evan-Moor Corp.

WEEK 19

Concept
Coins have an assigned value.

Counting Coins

Math Objective:
To help students identify coins and count different combinations of coins

Math Vocabulary:
cents, cent sign, coins, dime, nickel, penny, value, worth

Day 1 SKILLS

Number Sense
- Determine the value of a group of coins
- Count by 2s, 5s, or 10s

Literacy

Oral Language Development
- Respond orally to simple questions
- Use mathematical terms

Comprehension
- Make connections using illustrations, prior knowledge, or real-life experiences
- Answer questions about key details in a text read aloud

Introducing the Concept

Give each student 6 pennies and 2 nickels. Have students separate the two kinds of coins. Call students' attention to their pennies and say:

*How much is a **penny** worth?* (1 cent) *To count pennies, count by **ones**. Count your pennies. What is their **value**, or worth?* (6¢)

Write the number and ¢ symbol as you say:

*To show six cents, we write the number **6** and the **cent sign**: **6¢**. How much is a **nickel** worth?* (5 cents) *To count nickels, skip count by **fives**. Count your nickels. What is their value?* (10¢)

Write the number and ¢ symbol as you say:

*To show ten cents, we write **10¢**. Now count how much you have in all. Count the nickels first: **5, 10**. Now count on by ones: **11, 12, 13, 14, 15, 16**. What is the total value of your coins?* (16¢)

Listening to the Story

Distribute the Day 1 activity. Say: *Look at the picture and listen as I read a story about a boy who counts his grandpa's coins.*

Last week, Grandpa gave me **4** pennies and **3** nickels. I counted the nickels first: **5, 10, 15**. Then I counted on with the pennies: **16, 17, 18, 19**. It was 19¢. Today, Grandpa held out some coins. He said, "Guess how much money this is." I guessed 13¢. He handed me **5** nickels and **2** pennies. "This must be more than 13¢," I said.

Confirming Understanding

Distribute pencils. Develop the math concept by asking questions about the story. Say:

- Make an **X** on each nickel as you count it: **5, 10, 15, 20, 25**. How much are the nickels worth altogether? (25¢)
- Count on from **25** for each penny. (26, 27) How much money did Grandpa give the boy today? (27¢) Write **27¢** on the line.

Day 1 picture

© Evan-Moor Corp. • EMC 3039 • **Everyday Literacy: Math**

Week 19

Day 2 SKILLS

Number Sense
- Determine the value of a group of coins
- Count by 2s, 5s, or 10s

Literacy

Oral Language Development
- Respond orally to simple questions
- Use mathematical terms

Comprehension
- Make connections using illustrations, prior knowledge, or real-life experiences
- Answer questions about key details in a text read aloud

Reinforcing the Concept

Reread the Day 1 story. Then reinforce this week's math concept by discussing the story. Say:

The boy in our story counted coins. What coins did he count? (nickels and pennies)

Distribute the Day 2 activity and pencils. Say:

- *Look at row 1. It shows all pennies, so we will count by ones. Point to the first penny and start counting:* **1, 2, 3, 4.** *How much are these pennies worth?* (4¢) *Write* **4** *on the line. There is* **4¢** *in all.*

- *Look at row 2. It shows all nickels, so we will count by fives. Point to the first nickel and start counting:* **5, 10, 15, 20.** *How much are these nickels worth?* (20¢) *Write* **20** *on the line. There is* **20¢** *in all.*

- *Look at row 3. It shows nickels and pennies. Point to the first nickel and count by fives:* **5, 10.** *Now count on by ones:* **11, 12.** *How much are these coins worth?* (12¢) *Write* **12** *on the line. There is* **12¢** *in all.*

- *Look at row 4. It shows nickels and a penny. Point to the first nickel and count by fives:* **5, 10, 15.** *Now count on by ones:* **16.** *How much are these coins worth?* (16¢) *Write* **16** *on the line. There is* **16¢** *in all.*

Day 2 activity

Day 3 SKILLS

Number Sense
- Determine the value of a group of coins
- Count by 2s, 5s, or 10s

Literacy

Oral Language Development
- Respond orally to simple questions
- Use mathematical terms

Comprehension
- Make connections using illustrations, prior knowledge, or real-life experiences

Applying the Concept

Distribute the Day 3 activity and crayons. Then introduce the activity by reviewing pennies and nickels and introducing dimes:

A penny is worth 1 cent. A nickel is worth 5 cents. A dime is worth 10 cents.

- *Point to box 1. This candy costs* **3¢.** *How many pennies do you need to buy it?* (3) *Yes, 3 pennies equal 3¢. Color 3 pennies.*

- *Point to box 2. This lemon costs* **15¢.** *How many nickels do you need to buy it? Count on to find out. Stop when you get to 15:* **5, 10, 15.** *How many nickels did you count?* (3) *Yes, 3 nickels equal 15¢. Color 3 nickels.*

- *Point to box 3. This pencil costs* **20¢.** *There are different types of coins here. Let's see which ones will equal 20¢. Start counting at the dime:* **10.** *Count on by fives:* **15, 20.** *What coins did you count?* (dime, nickel, nickel) *Color 1 dime and 2 nickels. These coins equal 20¢.*

- *Point to box 4. This lollipop costs* **35¢.** *Let's see which coins we need to buy it. Start counting at the dime:* **10, 20.** *Now count on by fives:* **25, 30, 35.** *What coins did you count?* (dime, dime, nickel, nickel, nickel) *Color these coins, which equal 35¢.*

Day 3 activity

Day 4
SKILLS

Number Sense
- Determine the value of a group of coins
- Count by 2s, 5s, or 10s

Literacy

Oral Language Development
- Respond orally to simple questions
- Use mathematical terms

Comprehension
- Make connections using illustrations, prior knowledge, or real-life experiences

Extending the Concept

Distribute the Day 4 activity and pencils. Guide students through the activity by saying:

- *Look at the first group of coins. Let's count how much you have. Start with the dime: **10**. Count on by fives: **15, 20**. Count on by ones: **21, 22, 23**. You have **23¢**. Finish the sentence by writing **23** on the line.*

- *You buy candy that costs **3¢**. Cross out 3 pennies. Count the coins that are left: **10, 15, 20**. How much do you have left? (**20¢**) Write **20** on the line.*

- *So you had **23¢**. You spent **3¢**. How much money do you have left? Write a subtraction sentence that tells this story. Start by writing the amount you started with: **23**. Then, since you are subtracting, write a minus sign. Then write the cost of the candy: **3**. Write an equal sign. Then write how much you have left: **20**.*

Repeat the above steps with the remaining item.

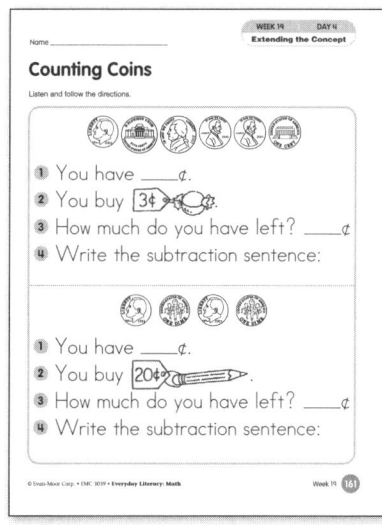

Day 4 activity

Day 5
SKILLS

Number Sense
- Determine the value of a group of coins

Mathematical Thinking and Reasoning
- Use number concepts for a meaningful purpose

Home–School Connection p. 162
Spanish version available (see p. 2)

Circle Time Math Activity

Reinforce this week's math concept with the following circle time activity:

Materials: pennies, nickels, and dimes (real or plastic); plastic reclosable bags

Preparation: Prepare individual "money bags" by placing a few coins in plastic bags. Make one bag for each student. Place all the money bags in a basket or box.

Activity: Gather students in a circle. Pass the basket around the circle and have each student take a money bag.

Ask students to open their bags and count their coins. Ask questions such as, *Does anyone have 21¢? Does anyone have 3 dimes? Does anyone have all pennies? Why are five nickels worth more than 5 pennies?* (Five nickels are worth 25¢ but 5 pennies are worth only 5¢.)

© Evan-Moor Corp. • EMC 3039 • *Everyday Literacy: Math* Week 19

Name _____

WEEK 19 | DAY 1
Confirming Understanding

Counting Coins

_____

158 Week 19

Everyday Literacy: Math • EMC 3039 • © Evan-Moor Corp.

Name _____

WEEK 19 | DAY 2
Reinforcing the Concept

Counting Coins

Skip count.

1 penny penny penny penny	___¢ in all
2 nickel nickel nickel nickel	___¢ in all
3 nickel nickel penny penny	___¢ in all
4 nickel nickel nickel penny	___¢ in all

© Evan-Moor Corp. • EMC 3039 • *Everyday Literacy: Math* Week 19

Name _____

WEEK 19 | DAY 3
Applying the Concept

Counting Coins

Color the coins.

1

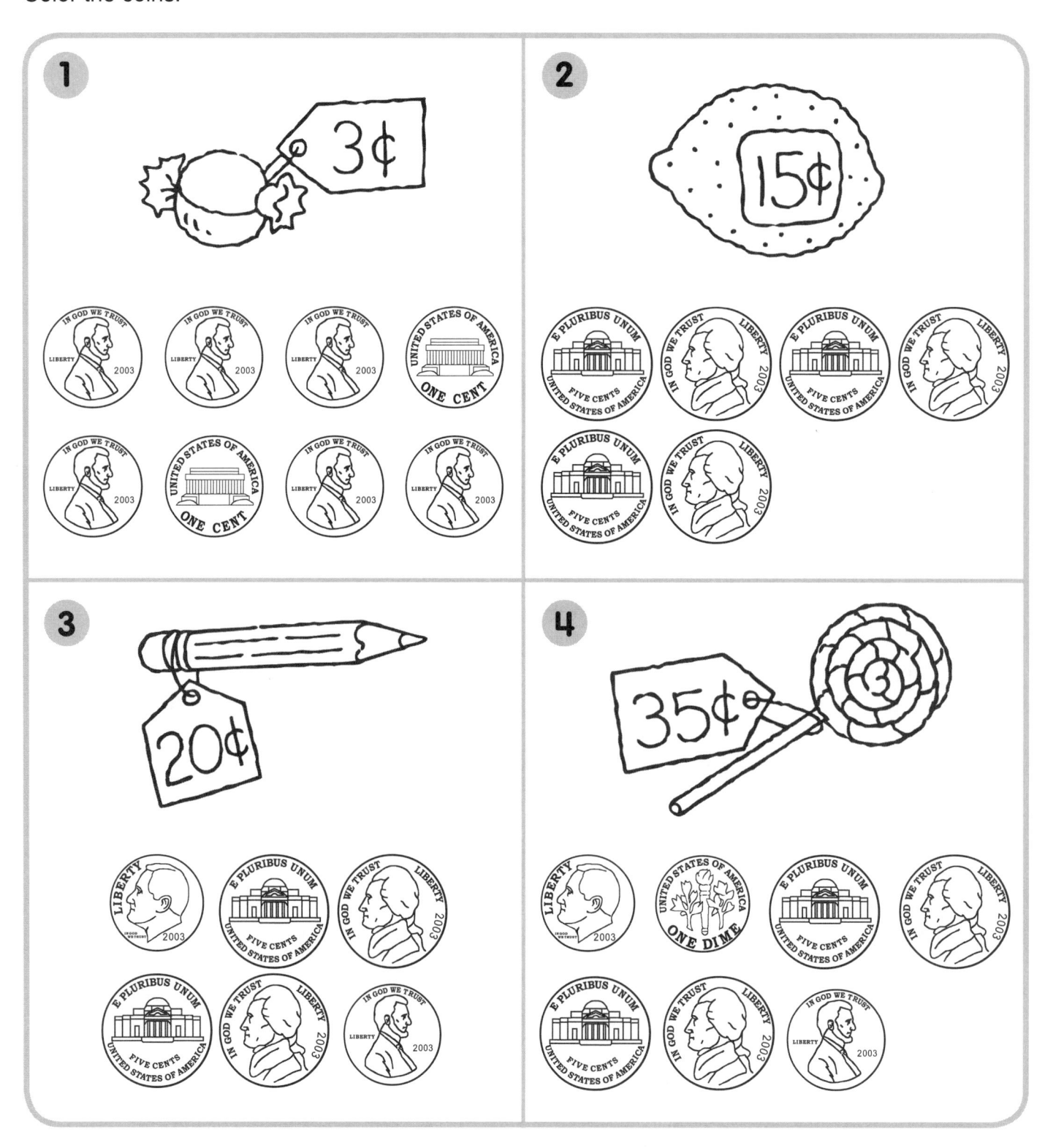

3¢

2

15¢

3

20¢

4

35¢

160 Week 19

Everyday Literacy: Math • EMC 3039 • © Evan-Moor Corp.

Name _____

WEEK 19 | DAY 4
Extending the Concept

Counting Coins

Listen and follow the directions.

1. You have ____¢.
2. You buy [3¢ 🍬].
3. How much do you have left? ____¢
4. Write the subtraction sentence:

---

1. You have ____¢.
2. You buy [20¢ ✏️].
3. How much do you have left? ____¢
4. Write the subtraction sentence:

© Evan-Moor Corp. • EMC 3039 • Everyday Literacy: Math

Name _____

What I Learned

What to Do
Have your child look at the pictures below and tell you how much each item costs. Then have your child color the coins needed to purchase each item.

WEEK 19

Home–School Connection

Math Concept: Coins have an assigned value.

To Parents
This week your child worked with dimes, nickels, and pennies.

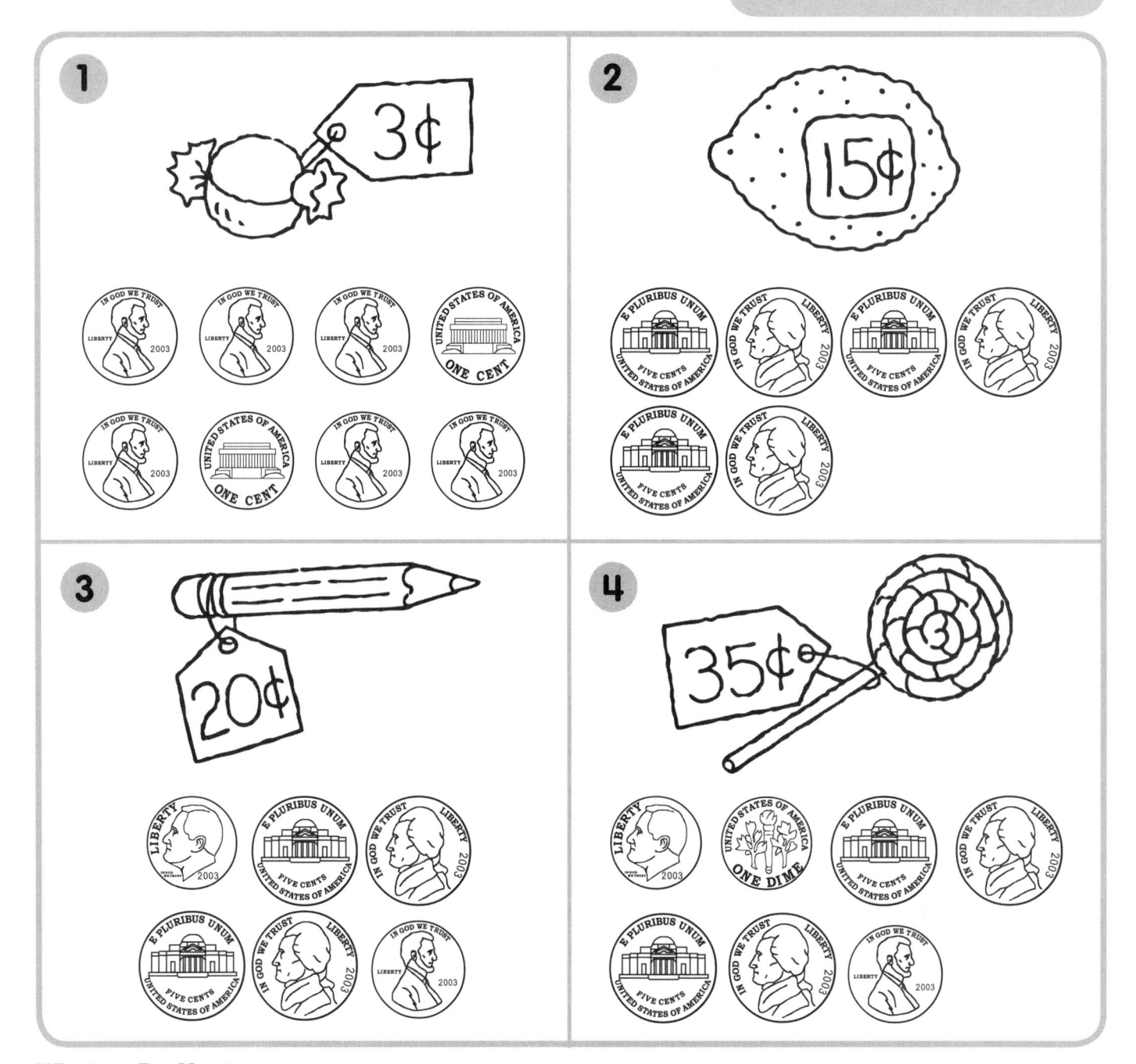

1. 3¢

2. 15¢

3. 20¢

4. 35¢

What to Do Next
Set a group of pennies, nickels, and dimes on a table. Invite your child to take a few, and you take the remaining coins. Each of you count the value of your coins and determine who has the most money.

WEEK 20

Concept
Data can be organized on a graph.

Using Graphs

Math Objective:
To help students organize, represent, and interpret data

Math Vocabulary:
bar graph, row, tally

Day 1 SKILLS

Data Analysis
- Organize, record, and interpret data

Literacy

Oral Language Development
- Respond orally to simple questions
- Use mathematical terms

Comprehension
- Make connections using illustrations, prior knowledge, or real-life experiences
- Answer questions about key details in a text read aloud
- Make inferences and draw conclusions

Introducing the Concept

Before the lesson, display a bar graph of 5 rows labeled **red**, **blue**, **purple**, **orange**, **green**. Title the graph **Favorite Colors**. Then distribute a self-adhesive note and crayons to each student. Refer to the graph as you say:

- *I have a question: What is your favorite color? Here are your choices: **red**, **blue**, **purple**, **orange**, **green**. On your paper, draw a shape. Color it your favorite color.* (students respond)

- *We will make a **bar graph** that shows how many children chose each color. When I say your name, come to the graph and place your sticky note in the row that shows your favorite color.* (students respond)

- *What is the longest **bar**, or **row**?* (students respond) *Which color do the most children like? Which is the least favorite?* (students respond)

Listening to the Story

Distribute the Day 1 activity. Say: *Look at the picture and listen as I read a story about a girl who makes a graph.*

---

*Tula asked 10 friends, "Which would you want as a pet: a rabbit, a monkey, or a squirrel?" Tula made a bar graph to show her friends' answers. She made **3** rows and gave each row a name: **rabbit**, **monkey**, **squirrel**. Then she colored a box for each friend's answer. "Hmm," said Tula. "I wonder if Mom would let me have a squirrel."*

Confirming Understanding

Develop the math concept by asking questions about the story. Ask:

- *What information does Tula's bar graph show?* (how many friends prefer a rabbit, monkey, or squirrel)

- *Why did Tula make 3 rows?* (one row for each answer choice)

- *Did more friends choose rabbit or squirrel?* (squirrel)

- *How many friends chose monkey?* (2)

Day 1 picture

Day 2 SKILLS

Data Analysis
- Organize, record, and interpret data

Literacy

Oral Language Development
- Respond orally to simple questions
- Use mathematical terms

Comprehension
- Make connections using illustrations, prior knowledge, or real-life experiences

Reinforcing the Concept

Reread the Day 1 story. Then reinforce this week's math concept by discussing the story. Say:

*The girl in our story made a **bar graph**. Let's use her graph to solve some problems.*

Distribute the Day 2 activity and pencils. Ask:

- **How many friends in all want a rabbit and a monkey?** *For number 1, we need to find out how many friends want a rabbit and a monkey. Look at Tula's graph at the top. First, write the number of friends who want a rabbit.* (3) *Next to the plus sign, write the number of friends who want a monkey.* (2) *Add: **3 + 2 = 5**. Write the sum, **5**, below the line.*

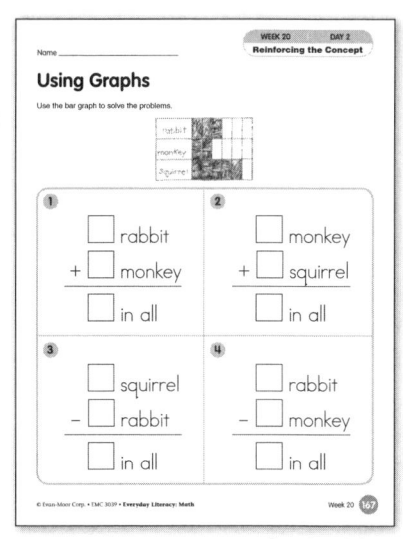

Repeat these steps for number 2, adding those who want a monkey and a squirrel.

- *Here is another problem:* **How many more friends want a squirrel than a rabbit?** *Will we add or subtract?* (subtract) *That's right. Let's write a subtraction sentence to tell this story. Point to number 3. First, write the number of friends who want a squirrel.* (5) *After the minus sign, write the number of friends who want a rabbit.* (3) *Now subtract: **5 − 3 = 2**. Write the difference, **2**, below the line.*

Repeat the process for number 4, finding how many more friends want a rabbit than a monkey.

Day 2 activity

Day 3 SKILLS

Data Analysis
- Organize, record, and interpret data

Literacy

Oral Language Development
- Respond orally to simple questions
- Use mathematical terms

Comprehension
- Make connections using illustrations, prior knowledge, or real-life experiences

Applying the Concept

Introduce the activity by describing a tally chart. Model making tally marks and say:

*We learned about **bar graphs**. Here is a different type of graph that uses tally marks. A **tally mark** is an up-and-down line. We draw one line for each item counted. For the fifth tally, we draw a line across. That makes it easy to count the sets of tallies by fives.*

Let's make a tally chart about paintbrushes, crayons, and markers.

Distribute the Day 3 activity and crayons. Say:

- *First, let's show how many paintbrushes there are. Cross out one paintbrush from the group. In the paintbrush row of the graph, draw one tally mark. Cross out another paintbrush above, then draw another tally mark in the same row. Keep going, crossing out paintbrushes and drawing tally marks. Don't forget to draw a line across for the fifth tally.*

- *How many tally marks are in the paintbrush row?* (6) *That's right. You have one set of **5** tally marks, plus **1** more. That makes **6**.*

Repeat the process with the remaining rows on the chart, guiding students in tallying 5 crayons and 3 markers.

Day 3 activity

Week 20   Everyday Literacy: Math • EMC 3039 • © Evan-Moor Corp.

Day 4 SKILLS

Data Analysis
- Organize, record, and interpret data

Literacy

Oral Language Development
- Respond orally to simple questions
- Use mathematical terms

Comprehension
- Make connections using illustrations, prior knowledge, or real-life experiences

Extending the Concept

Distribute the Day 4 activity and crayons. Then introduce the activity by saying:

Let's use the tally chart at the top of the page to make a bar graph. Then we'll use the bar graph to answer a question.

- *Look at the tally chart. Point to the paintbrush row. How many paintbrushes are tallied?* (7) *Show that information on the bar graph. Fill in **7** boxes of the paintbrush row.*

- *Look at the tally chart again. Point to the crayon row. How many crayons are tallied?* (8) *Show that information on the bar graph. Fill in **8** boxes of the crayon row.*

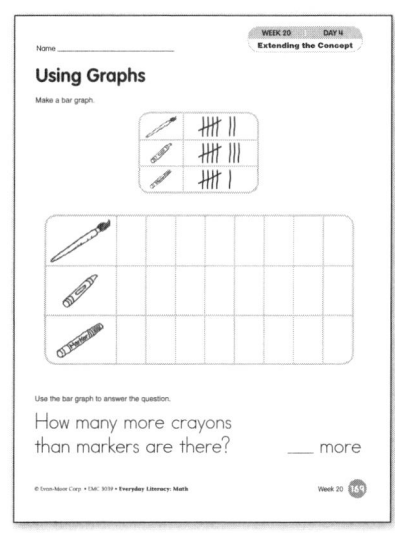

- *Look at the tally chart again. Point to the marker row. How many markers are tallied?* (6) *Show that information on the bar graph. Fill in **6** boxes of the marker row.*

- *Now let's use the bar graph to answer a question. Point to the question. Read along with me:* **How many more crayons than markers are there?** *In the space, use pictures, dots, or numbers to tell this story:* **8 – 6 = 2**. *What is the difference?* (2) *Write the number **2** on the line. There are **2** more crayons than markers.*

Day 4 activity

Day 5 SKILLS

Data Analysis
- Organize, record, and interpret data

Mathematical Thinking and Reasoning
- Use number concepts for a meaningful purpose

Home–School Connection p. 170
Spanish version available (see p. 2)

Hands-on Math Activity

Reinforce this week's math concept with the following hands-on activity:

Materials: number cards **1** through **30**, tally chart with two rows

Preparation: Make a set of number cards and a tally chart for each pair of students. Leave space on the left of each row for each student to write his or her name.

Activity: Divide students into pairs. Have each partner write his or her name on the graph. Then have students shuffle the cards and place them facedown in a pile.

Next, have each student take a card from the pile and compare their number with their partner's. Whoever has the greater number writes a tally mark in his or her row. Remind students that the fifth tally mark should be written across the previous four tally marks, and that they should count by fives.

Have students continue until all the cards are gone. Have volunteers say who had the most tally marks.

© Evan-Moor Corp. • EMC 3039 • **Everyday Literacy: Math** Week 20

Name _____

WEEK 20 | DAY 1
Confirming Understanding

Using Graphs

Name _____

WEEK 20 | DAY 2
Reinforcing the Concept

Using Graphs

Use the bar graph to solve the problems.

1

☐ rabbit
+ ☐ monkey
———
☐ in all

2

☐ monkey
+ ☐ squirrel
———
☐ in all

3

☐ squirrel
− ☐ rabbit
———
☐ in all

4

☐ rabbit
− ☐ monkey
———
☐ in all

© Evan-Moor Corp. • EMC 3039 • Everyday Literacy: Math
Week 20

Name _____

WEEK 20 | DAY 3
Applying the Concept

Using Graphs

Complete the tally chart.

Name _____

Using Graphs

Make a bar graph.

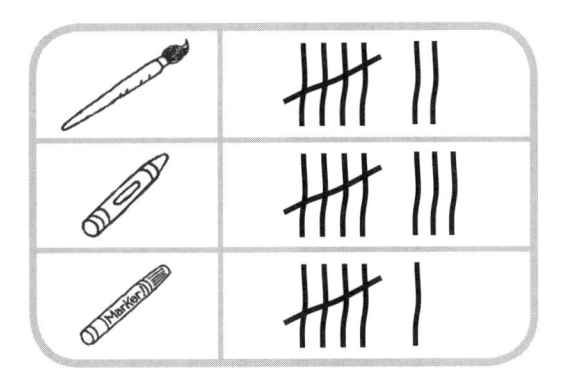

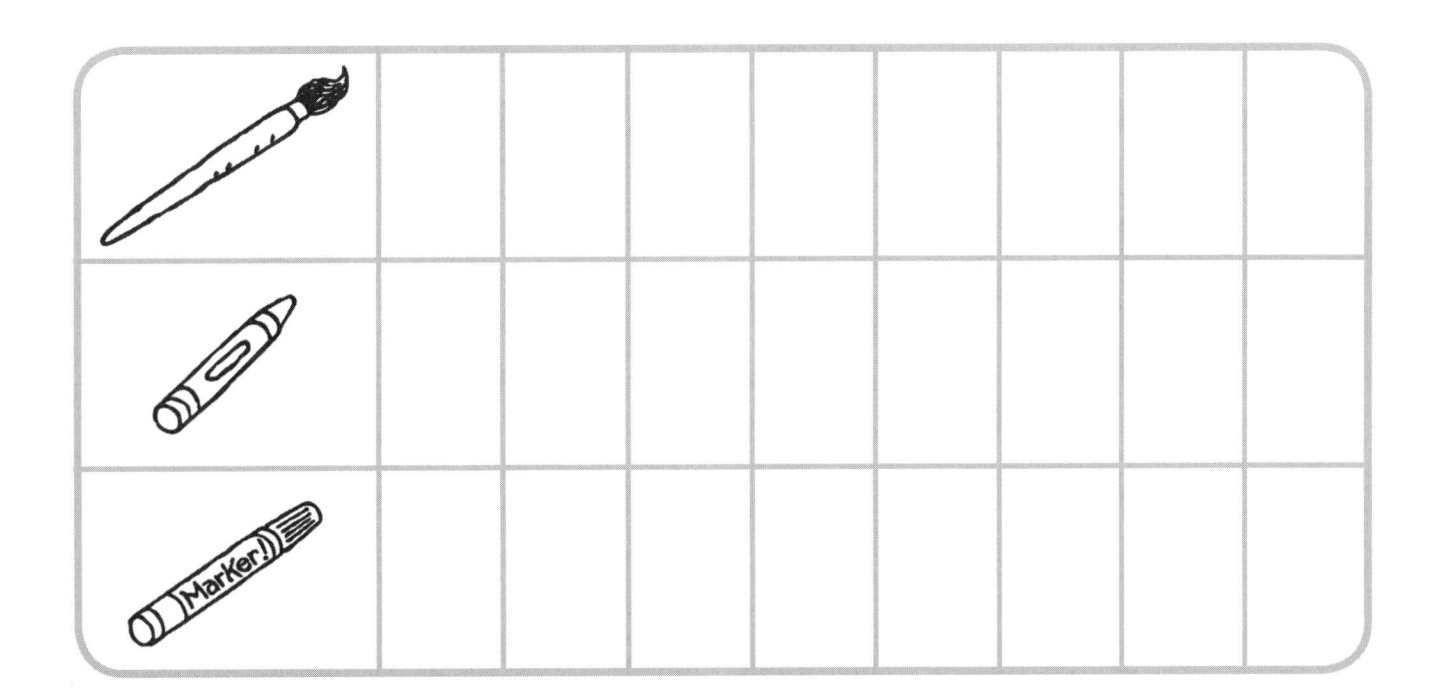

🖌							
🖍								
Marker						

Use the bar graph to answer the question.

How many more crayons than markers are there?   ___ more

WEEK 20 | DAY 4
Extending the Concept

Name _____

What I Learned

What to Do
Have your child look at the writing tools. Help your child count each paintbrush, crayon, and marker and then tally the results on the chart below. Ask your child questions such as, *How many more paintbrushes than markers are there?* (3)

WEEK 20

Home–School Connection

Math Concept: Data can be organized on a graph.

To Parents
This week your child learned to use bar graphs and tally charts.

What to Do Next
Set up a simple bar graph with the categories of paintbrushes, crayons, and markers. Help your child transfer the information from the tally chart above onto the bar graph.

170 Week 20

Everyday Literacy: Math • EMC 3039 • © Evan-Moor Corp.

Answer Key

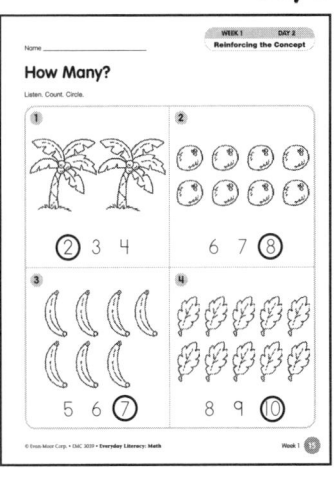

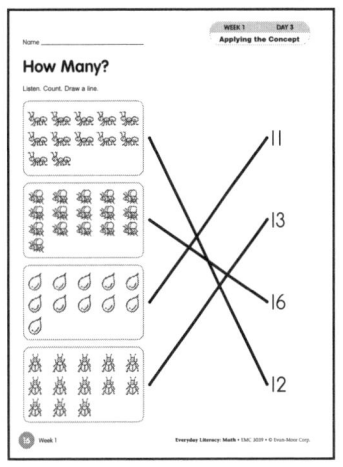

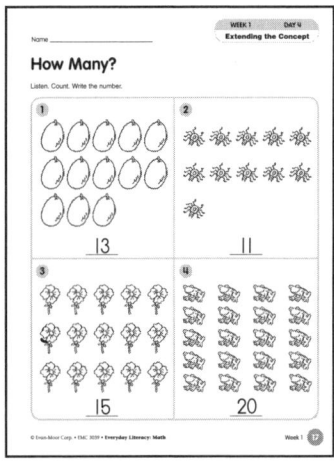

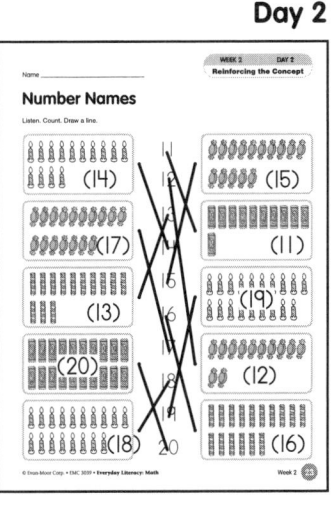

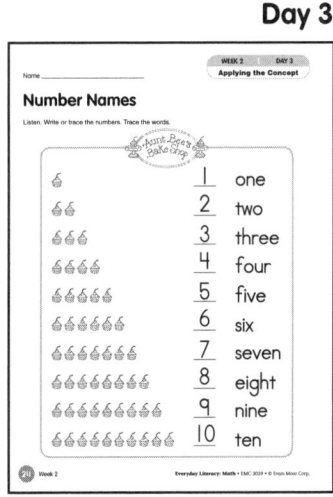

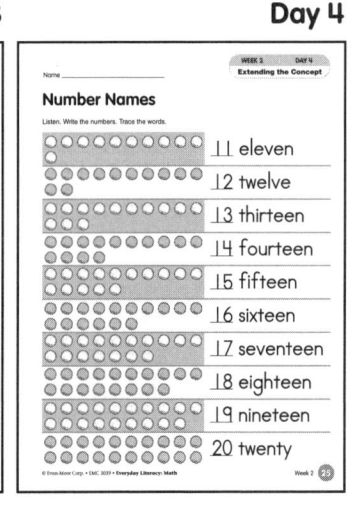

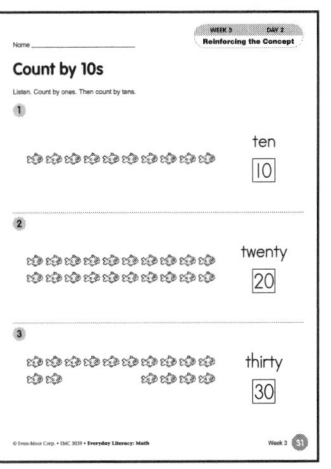

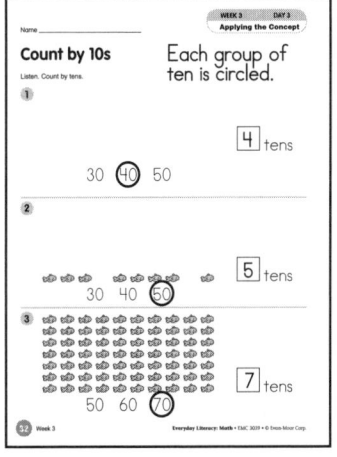

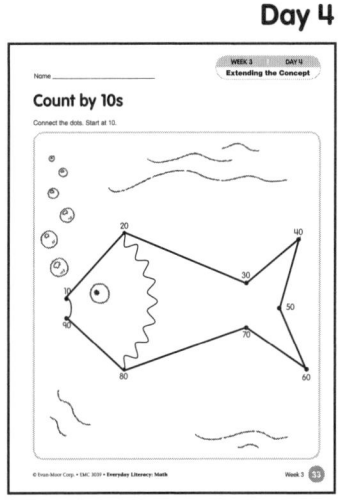

Week 1

Day 1 | Day 2 | Day 3 | Day 4

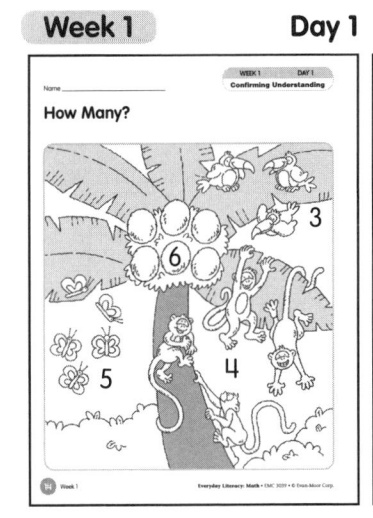

Week 2

Day 1 | Day 2 | Day 3 | Day 4

Week 3

Day 1 | Day 2 | Day 3 | Day 4

© Evan-Moor Corp. • EMC 3039 • Everyday Literacy: Math

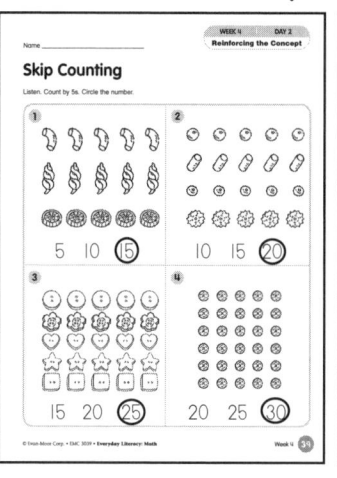

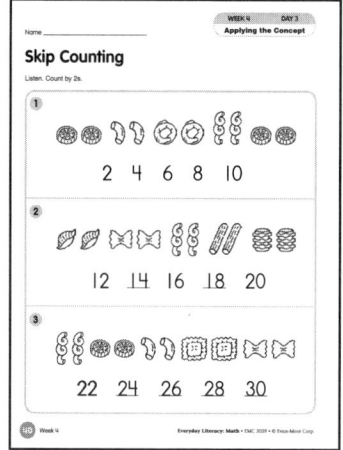

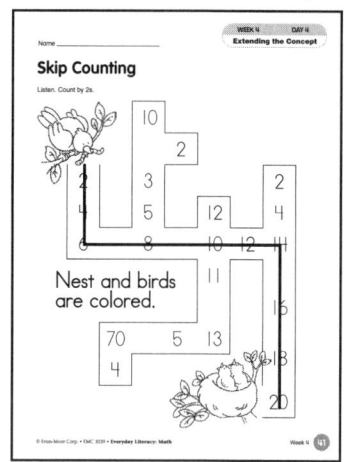

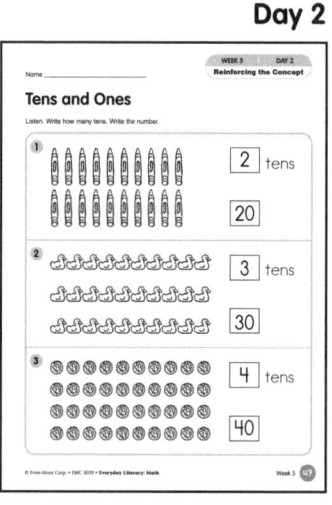

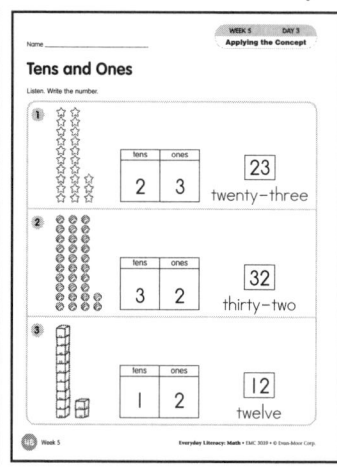

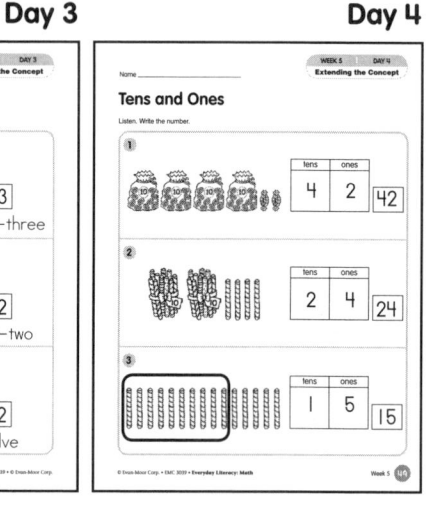

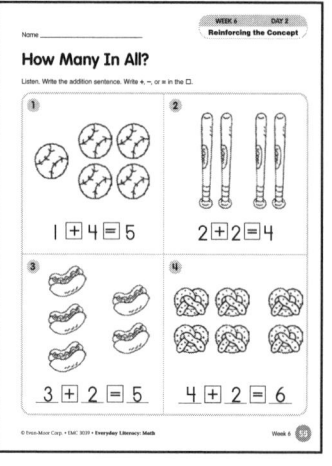

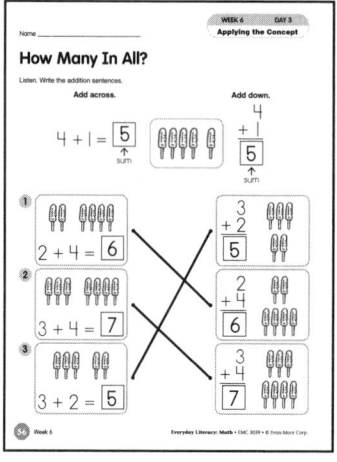

Week 4

Day 1 — Skip Counting — Confirming Understanding
Day 2 — Skip Counting — Reinforcing the Concept
Day 3 — Skip Counting — Applying the Concept
Day 4 — Skip Counting — Extending the Concept

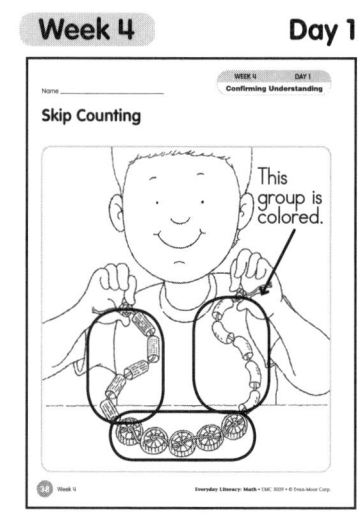

Week 5

Day 1 — Tens and Ones — Confirming Understanding
Day 2 — Tens and Ones — Reinforcing the Concept
Day 3 — Tens and Ones — Applying the Concept
Day 4 — Tens and Ones — Extending the Concept

Week 6

Day 1 — How Many In All? — Confirming Understanding
Day 2 — How Many In All? — Reinforcing the Concept
Day 3 — How Many In All? — Applying the Concept
Day 4 — How Many In All? — Extending the Concept

Everyday Literacy: Math • EMC 3039 • © Evan-Moor Corp.

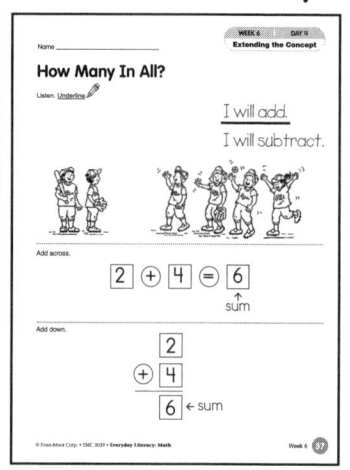

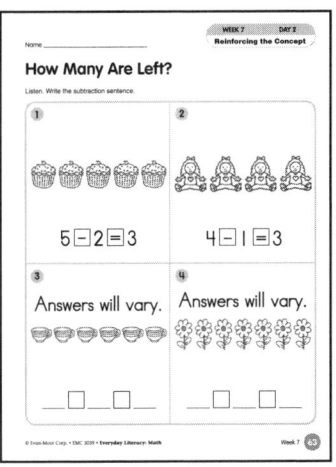

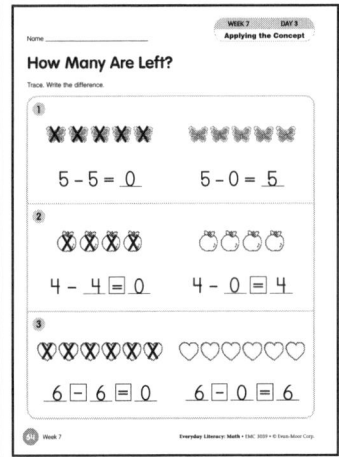

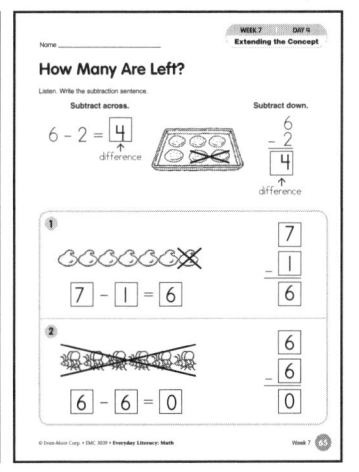

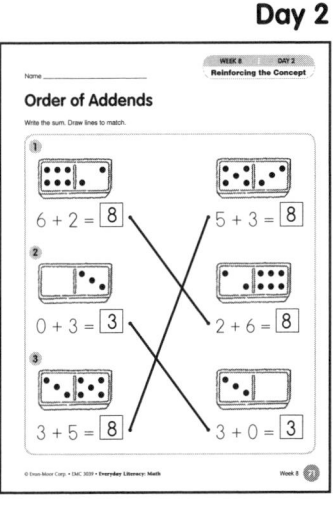

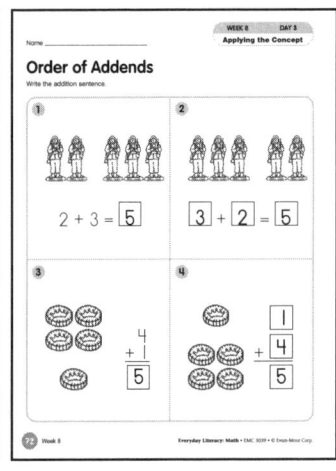

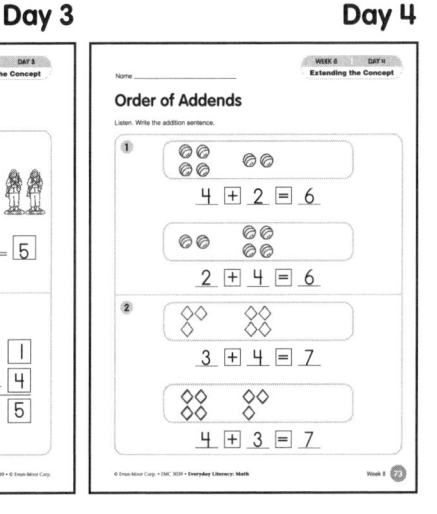

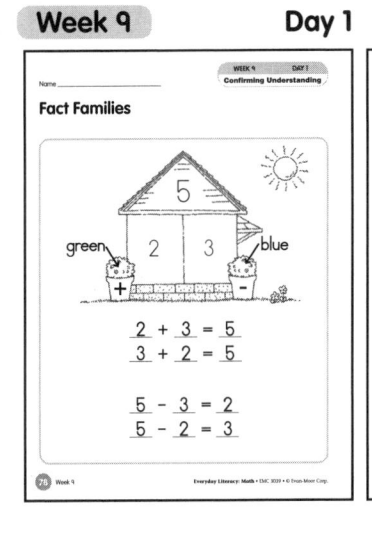

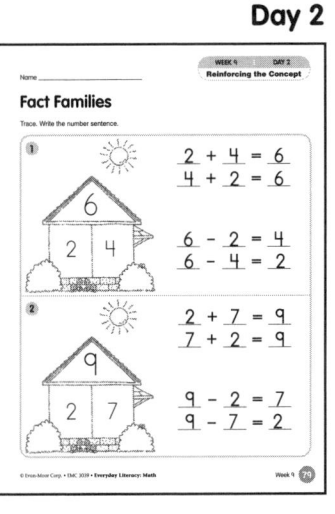

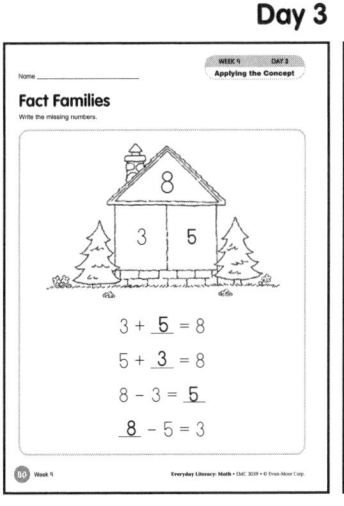

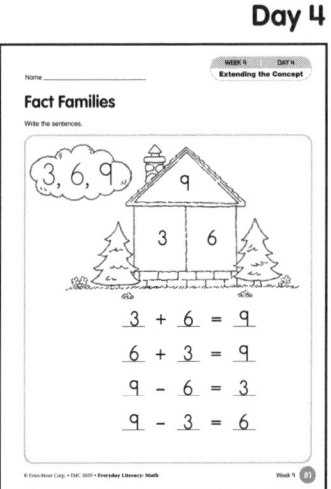

Week 7

Day 1 — How Many Are Left?
9 − 4 = 5

Day 2 — How Many Are Left?
1. 5 − 2 = 3
2. 4 − 1 = 3
3. Answers will vary.
4. Answers will vary.

Day 3 — How Many Are Left?
1. 5 − 5 = 0 ; 5 − 0 = 5
2. 4 − 4 = 0 ; 4 − 0 = 4
3. 6 − 6 = 0 ; 6 − 0 = 6

Day 4 — How Many Are Left?
6 − 2 = 4 (difference)
$\frac{6}{-2}/4$ difference

1. 7 − 1 = 6 ; $\frac{7}{-1}/6$
2. 6 − 6 = 0 ; $\frac{6}{-6}/0$

Week 8

Day 1 — Order of Addends
4 + 5 = 9   5 + 4 = 9

Day 2 — Order of Addends
1. 6 + 2 = 8 ; 5 + 3 = 8
2. 0 + 3 = 3 ; 2 + 6 = 8
3. 3 + 5 = 8 ; 3 + 0 = 3

Day 3 — Order of Addends
1. 2 + 3 = 5 ; 3 + 2 = 5
2. 4 + 1 = 5 ; 1 + 4 = 5

Day 4 — Order of Addends
1. 4 + 2 = 6 ; 2 + 4 = 6
2. 3 + 4 = 7 ; 4 + 3 = 7

Week 9

Day 1 — Fact Families
5 (green +, blue −) 2, 3
2 + 3 = 5
3 + 2 = 5
5 − 3 = 2
5 − 2 = 3

Day 2 — Fact Families
1. 6 ; 2, 4
2 + 4 = 6
4 + 2 = 6
6 − 2 = 4
6 − 4 = 2
2. 9 ; 2, 7
2 + 7 = 9
7 + 2 = 9
9 − 2 = 7
9 − 7 = 2

Day 3 — Fact Families
8 ; 3, 5
3 + 5 = 8
5 + 3 = 8
8 − 3 = 5
8 − 5 = 3

Day 4 — Fact Families
3, 6, 9 ; 9 ; 3, 6
3 + 6 = 9
6 + 3 = 9
9 − 6 = 3
9 − 3 = 6

© Evan-Moor Corp. • EMC 3039 • Everyday Literacy: Math

Week 10

Day 1
What's the Problem?

12, 13, 14, 15

12 + 3 = 15

Day 2
What's the Problem?
Count on to add.

1. 6, 7, 8
2. 4, 5, 6, 7
3. 5, 6, 7, 8, 9
4. 10, 11, 12, 13, 14, 15

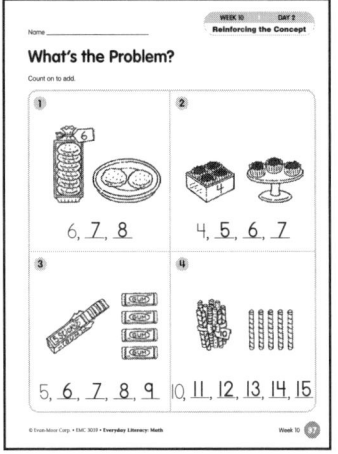

Day 3
What's the Problem?
Count back to subtract.

1. 5, 4, 3 — 5 − 2 = 3
2. 9, 8, 7, 6 — 9 − 3 = 6
3. 6, 5, 4 — 6 − 2 = 4
4. 7, 6, 5, 4 — 7 − 3 = 4

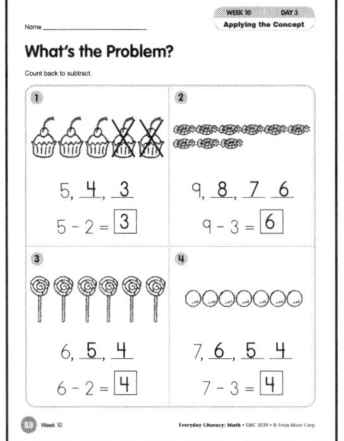

Day 4
What's the Problem?
Listen. Follow the directions.

1. 5, 6, 7, 8 — 5 + 3 = 8
2. 8, 7, 6, 5 — 8 − 3 = 5
3. 9, 8, 7, 6, 5 — 9 − 4 = 5
4. 6, 7, 8 — 6 + 2 = 8

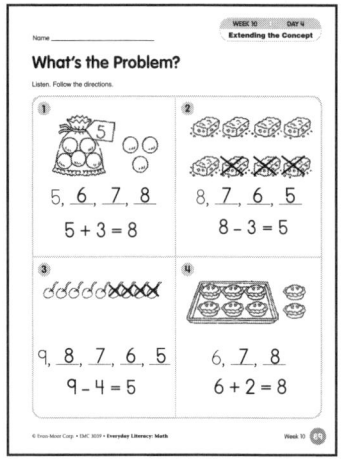

Week 11

Day 1
Greater Than, Less Than

23 < 25

Day 2
Greater Than, Less Than
Color the number that is greater.

1. 5, 4
2. 8, 6
3. 10, 13
4. 21, 12

Color the number that is less.

5. 3, 6
6. 7, 4
7. 9, 11
8. 20, 18

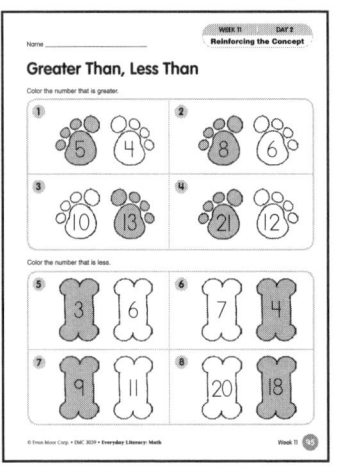

Day 3
Greater Than, Less Than
Circle >, <, or =.

1. 41 = 41
2. 24 < 26
3. 42 > 23
4. 30 < 40
5. 43 > 22
6. 16 = 16

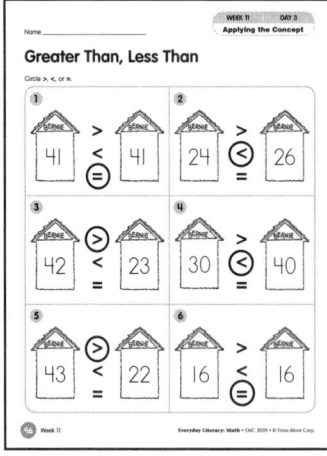

Day 4
Greater Than, Less Than
Count on. Write the number. Circle >, <, or =.

1. 8 < 9
2. 13 = 13
3. 12 > 9

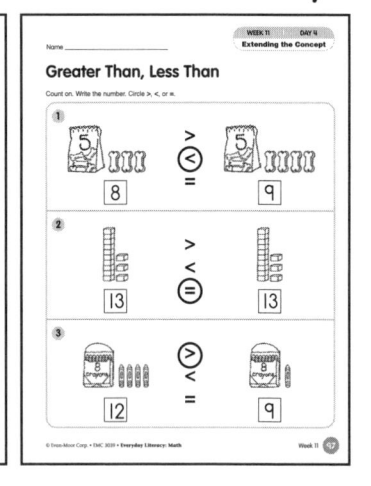

Week 12

Day 1
Basic Shapes

red, green, brown

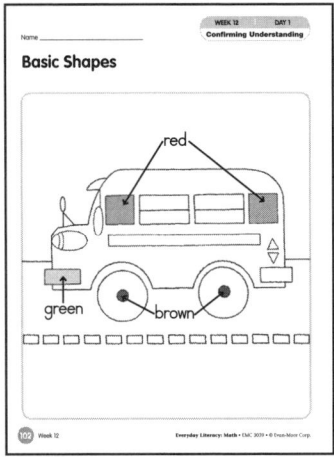

Day 2
Basic Shapes
Draw lines to match.

1. triangle
2. square
3. circle
4. rectangle

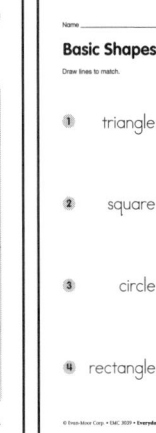

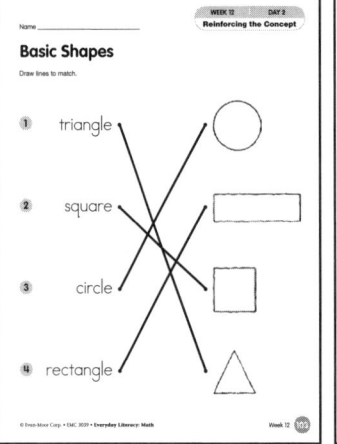

Day 3
Basic Shapes
Color.

1. No Corners
2. Equal sides
3. 3 Corners
4. 4 Corners

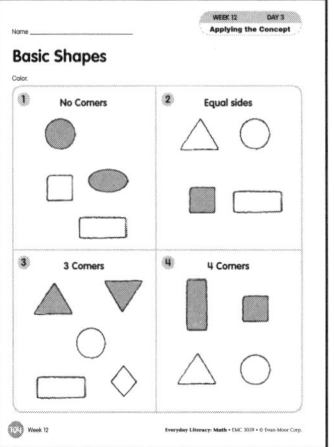

Day 4
Basic Shapes
Count.

1. 2 rectangles, 4 triangles
2. 2 triangles, 4 squares, 3 rectangles, 7 circles

174

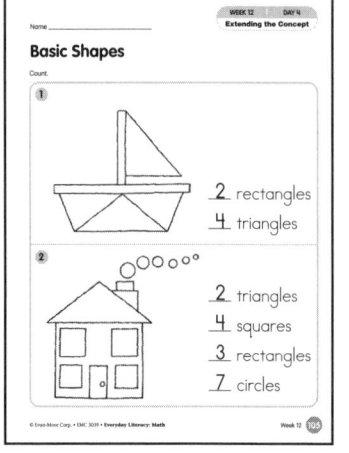

Everyday Literacy: Math • EMC 3039 • © Evan-Moor Corp.

Week 13

Day 1 — Fractions: Each half is colored a different color.

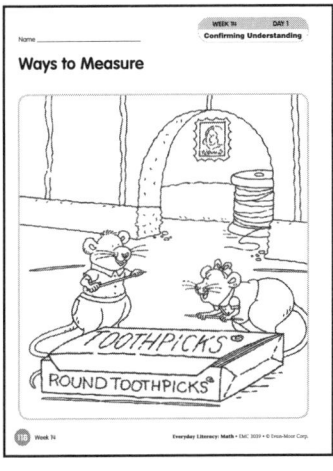

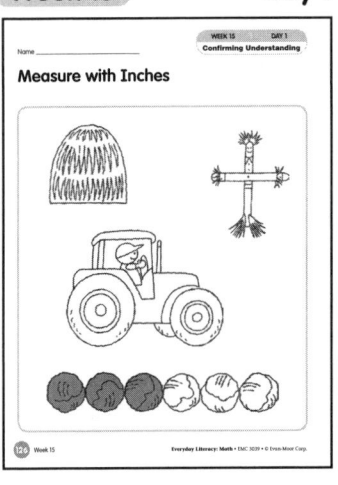

Day 2 — Fractions: Listen. Color the circle next to **yes** or **no**.
1. ● no
2. ● yes
3. (blank)
4. ● yes

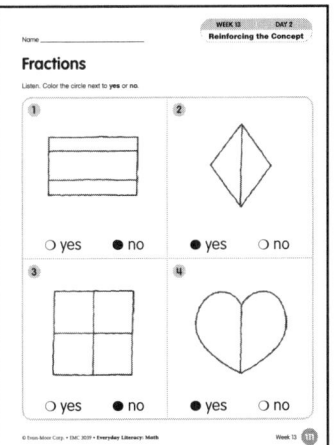

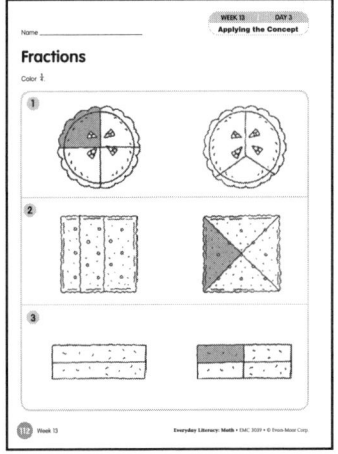

Day 3 — Fractions: Color ¼.

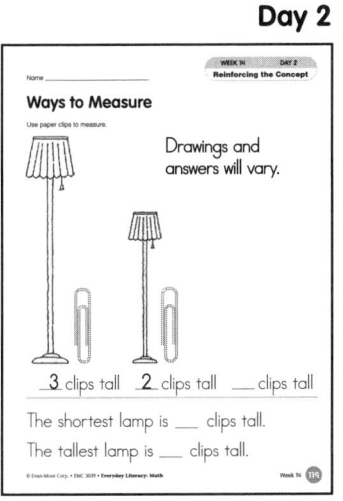

Day 4 — Fractions: Color one part. Circle the fraction.
1. ½ — ¼ (circled)
2. ½ — ¼
3. ½ — ¼ (circled)
4. ½

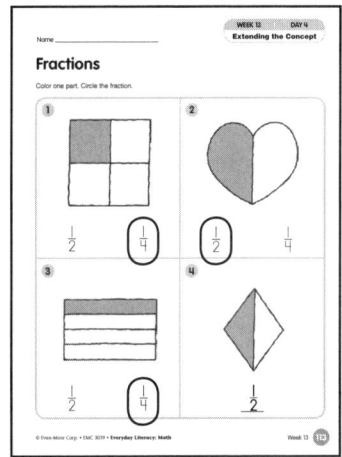

Week 14

Day 1 — Ways to Measure

Day 2 — Ways to Measure: Use paper clips to measure. Drawings and answers will vary.
_3_ clips tall   _2_ clips tall   ___ clips tall
The shortest lamp is ___ clips tall.
The tallest lamp is ___ clips tall.

Day 3 — Ways to Measure: Guess. Then measure with a paper clip.
guess ☐   measure 2
guess ☐   measure 3

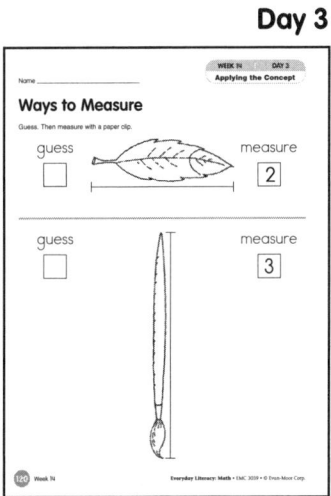

Day 4 — Ways to Measure: Color the items that are taller than one paper clip.

Week 15

Day 1 — Measure with Inches

Day 2 — Measure with Inches: Estimate. Measure. Estimates will vary.
estimate: ___ inches   measure: _3_ inches
estimate: ___ inches   measure: _2_ inches
estimate: ___ inches   measure: _4_ inches

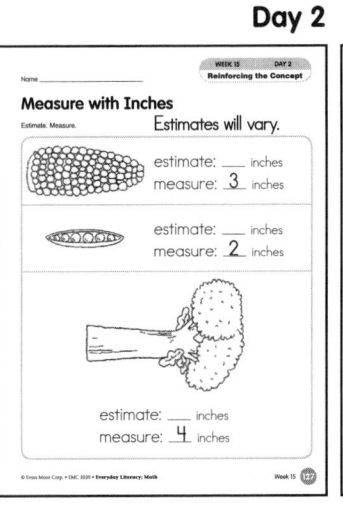

Day 3 — Measure with Inches: Estimate. Measure. Estimates will vary.

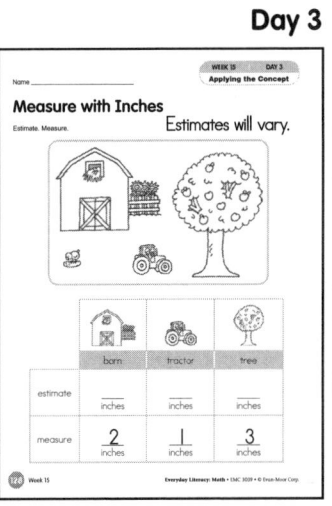

	barn	tractor	tree
estimate	inches	inches	inches
measure	2 inches	1 inches	3 inches

Day 4 — Measure with Inches: Measure.
2 inches, 4 inches, 1 inches, 5 inches

© Evan-Moor Corp. • EMC 3039 • Everyday Literacy: Math

Week 16

Day 1
Time for Fun!

Day 2
Time for Fun!
Write the time.
1. 10:00
2. 7:00
3. 3:00
4. 6:00

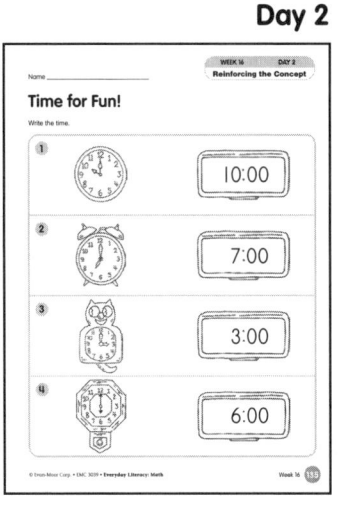

Day 3
Time for Fun!
Write the time.
1. 8:00 — 8 o'clock
2. 1:00 — 1 o'clock
3. 4:00 — 4 o'clock

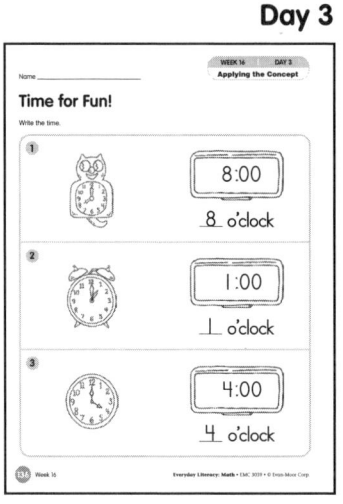

Day 4
Time for Fun!
Write the time.
1. 9:00
2. 2:30
3. 11:30
4. 11:00

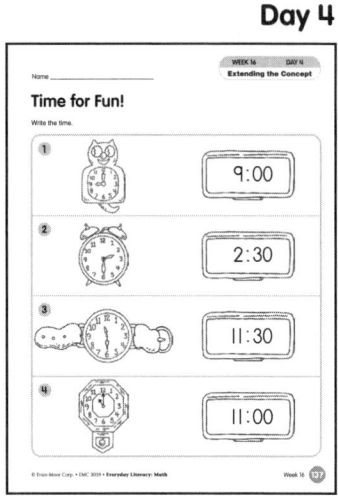

Week 17

Day 1
Patterns

Day 2
Patterns
Circle what comes next.

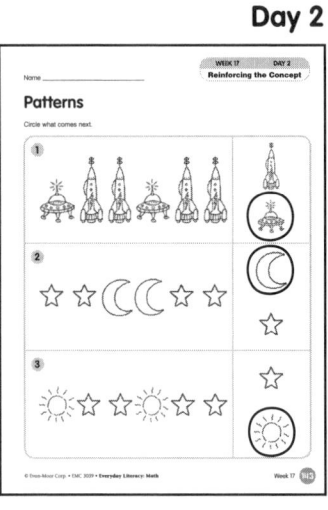

Day 3
Patterns
Show the pattern with A B.
1. A B A B A B
2. A A B A A B
3. A A B B A

Day 4
Patterns
Continue the number patterns.
1. 2, 4, 6, 8, 10, 12
2. 10, 20, 30, 40, 50, 60
3. 5, 10, 15, 20, 25, 30

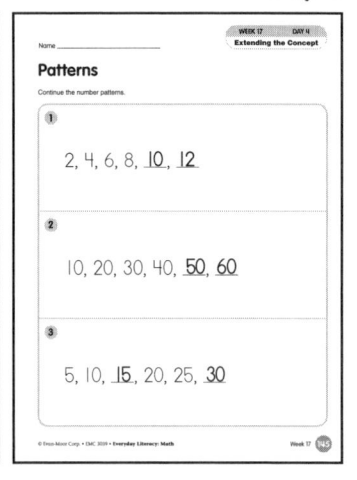

Week 18

Day 1
Where Is It?

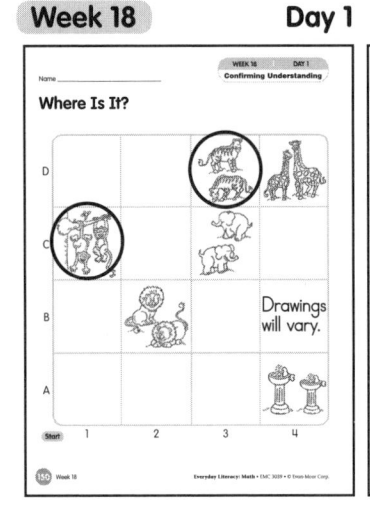

Day 2
Where Is It?
Listen and follow the directions.
Colored yellow.

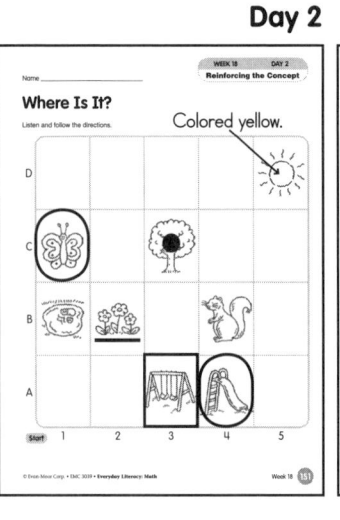

Day 3
Where Is It?
Listen and follow the directions.

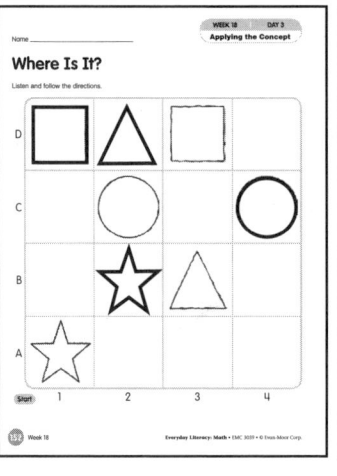

Day 4
Where Is It?
Listen. Color the squares.

	1	2	3	4	5
E	R	B	G	B	R
D	B	O	G	O	B
C	G	G	R	G	G
B	B	O	G	O	B
A	R	B	G	B	R

Key
Red: 1A, 1E, 3C, 5A, 5E
Orange: 2B, 2D, 4B, 4D
Blue: 1B, 1D, 2A, 2E, 4A, 4E, 5B, 5D
Green: 1C, 2C, 3A, 3B, 3D, 3E, 4C, 5C

Drawings will vary.

Everyday Literacy: Math • EMC 3039 • © Evan-Moor Corp.

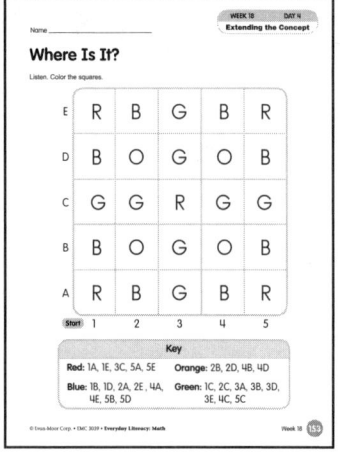